The Broken Hoe

DAVID URU IYAM

The Broken Hoe

CULTURAL RECONFIGURATION IN BIASE SOUTHEAST NIGERIA

University of Chicago Press
CHICAGO & LONDON

David Uru Iyam has taught at Emory University and at the University of California, Los Angeles, and has served as acting executive director of the African Studies Association.

The University of Chicago Press, Chicago 60637
The University of Chicago Press, Ltd., London

Printed in the United States of America

04 03 02 01 00 99 98 97 96 95 1 2 3 4 5

ISBN: 0-226-38848-4 (cloth)
ISBN: 0-226-38849-2 (paper)

Library of Congress Cataloging-in-Publication Data

Iyam, David Uru.
The broken hole : cultural reconfiguration in Biase southeast Nigeria / David Uru Iyam.
p. cm.
Includes bibliographical references (p.) and index.
1. Biase (African people)—Social conditions. 2. Biase (African people)—Economic conditions. 3. Biase (African people)—Agriculture. 4. Rural development—Nigeria—Cross River State. 5. Economic development—Nigeria—Cross River State. 6. Social change—Nigeria—Cross River State. 7. Cross River State (Nigeria)—Politics and government. 8. Cross River State (Nigeria)—Environmental conditions. I. Title.
DT515.45.B5I93 1995
307.72'09669'44—dc20 94-39003
CIP

♾The paper used in this publication meets the minimum requirements of the American National Standard for Information Sciences—Permanence of Paper for Printed Library materials, ANSI Z39.48-1984.

In remembrance of my grandmother

Aneji Jenny Ekeko

Contents

Illustrations

Figures

Tables

Preface

The ability of rural communities to manage drastic cultural changes has been weakened by unfamiliar coping strategies often erroneously perceived as indigenous to them. As technologically less complex cultures incorporate new cultural elements generated from within and outside the society, the resulting reconfiguration of indigenous institutions affects their communal management strategies, influences rural socioeconomic relations, and creates disadvantageous conditions for economic growth. Such a breakdown in a culture's traditional problem-solving tools results in the entire system malfunctioning so badly that it ceases to be what it was. In the process, current practices become maladaptive, in the face of present social realities, to economic growth. From this perspective I examine changes in the subsistence strategies of the rural Biase of southeast Nigeria. To do this, I examine the relationship between Biase indigenous coping strategies and new patterns of exploiting environmental resources and their consequences for rural livelihood. I suggest that the ongoing cultural changes in Nigeria's rural communities adversely affect efforts to change the economic conditions of rural communities using only their indigenous technology.

Earlier perspectives on Africans have been criticized as being Eurocentric because of their implication that existing African institutions are constraints rather than frameworks for development. While this criticism is correct, it is also idealistic and misplaced. Its idealism is partly a result of the misconception that current configurations of rural institutions and the practices and behaviors that sustain them are indigenous to the societies we study. A misplaced antagonism is therefore directed at cultural practices that are foreign to indigenous societies. The prevailing cultural practices of many groups have long since been significantly altered by pressures and constraints beyond the control of the group as well as by processes they are unable to control.

Some scholars understandably take a neutral stance when confronted with cultural practices that are not firmly embedded in mainstream criticism. Very often we rationalize certain practices within the cloak of cultural relativism. I am conscious of this in writing this book. I report on the Biase from my dual capacity as interpreter and interpreted. I present a view of my ethnic group that neither glamorizes nor castigates our cultural behaviors and practices but attempts to understand the reality of Africa's rural conditions. I interpret this reality as much as I am interpreted by it. The book treats topics typically covered in an introductory anthropology course, such as technology, economy, social organization, politics, and religion. This gives students a referent for comprehending cultural paradigms as well as a context for understanding a traditional African culture. I offer an ethnography of the Biase through the perspective of community members who explain their culture and suggest reasons for the weakening of their cultural practices. I examine how current practices and behaviors intrinsic to Biase communities have made their traditional conventions and institutions inoperative or dysfunctional.

In writing this book, I have benefited from the advise and suggestions of my teachers in the University of California, Los Angeles. Special thanks go to Teshome Gabriel, Richard Sklar, Francesca Bray, Nancy Levine, Robert Edgerton, and Peter Hammond for their critical comments on the first draft of the manuscript. I am particularly indebted to Peter Hammond, who nurtured my interest in anthropology and whose scholarly perspective encouraged my research. I am grateful to the Rockefeller Foundation for the research fellowship that made this study possible; their generous funding also enabled Peter Hammond to visit me in the field during my research. I would like to acknowledge my gratitude to my students and colleagues at Emory University who provided an intellectually stimulating yet relaxing environment that enabled me to complete this book. My greatest debt is to my daughters, Onne, Boma, and Ejeogba, who provided enough diversion to relieve me of the pressure of writing this book, and to my wife, Ruby, without whose optimism, devotion, and love I would not have contemplated this study.

ONE

Issues in Rural Development

As sub-Saharan Africa moves into the twenty-first century, concern with improvement of life quality will continue to be of great interest in the region. But the inability to devise effective solutions to such accumulated rural problems as inadequate health facilities, inadequate rural transportations, lack of surplus food production, an inadequate water supply, and a stagnant economy will persist; so also will the "expert" views that have, over the years, shaped various plans for resolving these problems. Although the split between those seeking to supplant existing technology (Bendix 1967; Crow et al. 1988; Kiwanuka 1986) and those seeking to achieve change by modifying or augmenting indigenous technology (Wey 1988) has been minimally mediated, these foci are still significant to various degrees as scholars and development agencies attempt to articulate the conditions of rural communities in developing economies.

A major concern with changing rural conditions, as expressed by some scholars, is fear of the growing destitution of villagers due to uncontrolled contact with modernity. Threats of a reversal of their "political independence, reliance on local natural resources, and relative internal social equality" (Bodley 1990: 3) are often expressed. According to Bodley (one of the foremost advocates for nonindustrial people) nonindustrial people have survived longer than industrial civilizations and would continue to manage their environment expertly if left undisturbed. However, rural people are not isolated. They are continuously exposed to elements of modernity and actively seek to live like city people. Rural people do not seek modernity out of frustration with traditional ways of life. Rather, they acknowledge their need to be more effective in the face of drastic cultural changes, as familiar coping strategies are weakened by new forms.

Bromley and Cernea (1989: 17) have suggested that

> in many cultures conformity with group norms at the local level is an effective sanction against antisocial behavior. A vi-

> able common property regime thus has a built-in structure of economic and non-economic incentives that encourages compliance with existing conventions and institutions. Unfortunately in many settings, those sanctions and incentives have become inoperative—or dysfunctional—largely because of pressures and forces beyond the control of the group, or because of internal processes that the group wasn't able to master.

If there is a breakdown in a people's system of rights and entitlement, including management and authority instruments, "the entire system malfunctions up to a degree at which it ceases to be what it was" (Bromley and Cernea 1989: 18). This breakdown is the consequence of colonialism, which redefined indigenous institutions; of new nation-states within which rural communities must continuously negotiate their legitimacy; and of competing neighboring groups with whom they are in continuous interaction. The result of this systemic interaction is a reconfiguration of the institutional and organizational arrangements of rural people, which significantly decreases the effectiveness of their traditional strategies for sustaining economic growth. My ethnic group, the Biase of southeastern Nigeria, provided important indicators for this suggestion.

In my study of the Biase I locate indigenous coping strategies within their historical context, and explore the relationship between current Biase institutional arrangements and their response to environmental resource exploitation, new patterns of social interaction, and rural economic growth. I aim to understand how our transformed cultural institutions influence the strategies we use to manage and exploit local and external resources, and how they affect economic growth. For instance, what effects do changing gender relations in market exchange have on our economy? How does division of labor by age and gender affect communal resource management? How does our weak polity affect the role and functions of local associations, our status with the state, and our ability to attract government economic assistance? And how do beliefs and ideologies influence resource management? Following Bromley and Cernea, I argue that (1) the capability of indigenous practices to manage rural social organization has been overwhelmed by the demands of new cultural elements generated from within and outside the society; (2) the resulting reconfiguration of indigenous institutions renders them less effective for communal management; (3) consequently, a weakened array of cultural conventions and institutions have become minimally able to influence rural socioeconomic relations and have created disadvantageous conditions for economic growth.

The "Myth of Merrie Africa"

Donham (1985) reminds us of A. G. Hopkins's parody of the "myth of Merrie Africa," a belief in a precolonial Golden Age "in which generations of Africans enjoyed congenial lives in well-integrated, smoothly functioning societies. The means of livelihood came easily to hand . . . and this good fortune enabled the inhabitants to concentrate on leisure pursuits, which, if some sources are to be believed, consisted of interminable dancing and drumming" (Hopkins, cited in Donham 1985: 11). When I arrived back among my people to begin my research, the sound of drumming echoed only from a distant past, and the community members seemed to be waiting uncertainly for the resumption of dancing. Part of the reason for this was that the most important strategies we used to manage our society a few decades ago have now become our most crippling institutions: our once-effective market economy is today weakened by lack of male participation; beliefs that encouraged conformity with social rules have been tested, violated, and scorned by the young.

A little over twenty years ago, the Biase of Nigeria's Cross River State (fig. 1) had a flourishing trade in yams, palm oil, fish, and vegetables. We firmly controlled the palm oil and yam trade along the Cross River coast, had few of the transportation difficulties our inland trade partners experienced, and sent patrol canoes to collect "water rent" from non-Biase groups, such as the Afikpo Igbo, for occasional use of the river (Nair 1972). Besides this commanding economic position, we also maintained powerful social institutions that were pivotal in ordering the behavior of members. Secular and mystical male and female associations commanded fear and reverence from members whose affiliations were deemed to have supernatural sanction. In some instances, allegiance to mystical associations tended to supersede kinship bonds; Biase villagers noted that good standing in mystical associations oftentimes depended on how many members of one's family had been supernormally sacrificed or "killed" for mystical purposes. Village elders who belonged to such associations had little difficulty in enforcing conformity with social rules or in directing the affairs of their communities.

Today, the Biase face social and economic pressures which exert significant stress on their indigenous institutions, reconfiguring them

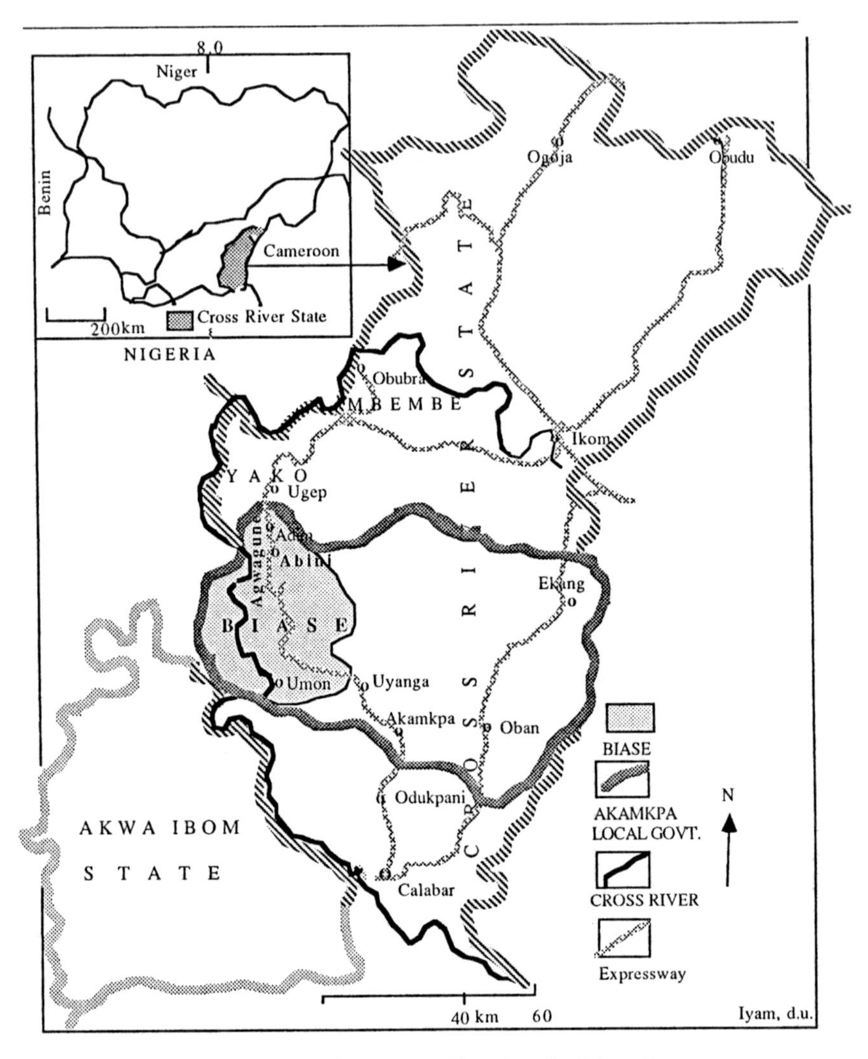

FIGURE 1. Cross River State, Showing the Biase Group

into new forms that are minimally effective for coping with present-day realities. This is evident in the dilemma we face as we attempt to manage our environment, make use of our technology, sustain our economy, and negotiate our social and political relations.

Cultural changes contribute to the accelerated rate of environmental deterioration, particularly the problem of population growth. But added to this is the greater socioeconomic independence of Biase youths, which has lowered the age at which young men and women must fend for themselves and has increased the pressure on environmental resources. Plants and animals take longer to find; hunting is no longer as easy and profitable as it used to be. Today hunters travel three hours longer to reach hunting areas and spend more time searching for game. Intergroup conflict has become more deadly because spears and bows and arrows have been replaced by guns. Consensual and elected authorities are no longer effective leaders, as they are now on the state payroll and are more accountable to the central government than to their constituents.

Novel social arrangements affect previously stable units in new ways. Although the need to assure a large labor pool through polygamy and a high birthrate persists, the state's mandate to attend school and government campaigns discouraging large families have lessened the economic importance of children. Where the birthrate remains high, it is more for parental prestige, the perception of wealth, or security in old age than for fulfilling any immediate economic need.

Beliefs and practices that formerly threatened retribution and served as guides to acceptable behavior have been tested and demystified. The result is a rejection of the idea of supernatural retribution, disrespect for community elders, and disregard for social conventions. With the disappearance of many of the cultural forms that made us effective managers of our environment, we have adopted new techniques of cultural and environmental management. Practices that once made us efficient managers of our natural setting are now maladaptive for economic growth.

Reconfiguration and the Weakening of Cultural Institutions

Unclear about the processes that have decreased the efficiency of our traditional institutions, we have attempted to cope with our dilemma by restructuring some of our social conventions and behaviors; this strategy is common among other Nigerian communities.

In 1990 the traditional ruler of the southeastern town of Calabar banned the celebration of a favorite festival, "Agaba," because it put lives and property in danger.

Women's associations, in particular the Nigerian Organization of Women, have in the past fifteen years embarked on a reassessment of the status of women. In 1987, *The African Guardian,* after doing a survey on marriage trends in Nigeria, reported a reversal of indigenous marriage arrangements. Contrary to traditional practices, men were increasingly looking for educated wives who would supplement their income. Women too preferred an arrangement that would make it possible for them to work outside the farm. Although large families are said to be conducive to the high labor requirement in less-developed economies, Biase women enthusiastically welcomed a government suggestion that they bear no more than four children. A women's group suggested that Nigerian women should start marrying under statutory laws which recognize their rights to spousal inheritance, rather than customary laws, which deprived them of inheritance at the death of their husbands (*African Guardian*, April 29, 1991: 36). There is a strong movement among educated Nigerian women to discourage the practice of clitoridectomy among some Nigerian groups, which is alleged to be done unhygienically and to be life-threatening.

Traditional medicine men now license their trade with the government as a check against fraudulent practitioners. Ritual observances once perceived as the lifeline of certain communities have been either secularized or banned by the indigenous people themselves, who see those practices as socially constraining within present realities; these observances include the Egup association of Agwagune, the Ekpirikpe dance of Ohafia, and the Egugun festival of Lagos. In the middle seventies, the Nigerian government banned civil servants from membership in some traditional associations which the government perceived as heightening the incidence of favoritism in access to public office.

Chieftaincy disputes in Biase and most other Nigerian groups suggest that homogeneity no longer guarantees decision by consensus; factional interests threaten unity. Many of these changes result from local and foreign influences that have left in place only skeletal images of once-effective indigenous practices and institutions. At the same time, these cultural vestiges have been accepted as indigenous both by those of us who practice them and by development agencies and their consultants. Where no vestiges are left, old and forgotten practices and behaviors are resurrected to contest present-day realities.

Problems in Rural Development

The literature on development often focuses more on the reasons for development failures than for successes. Critics of Nigeria's rural conditions probably have less to apologize for than critics of urban development. The Biase experiences indicate that rural communities are less likely to be specifically targeted for development initiatives than cities; projects started in villages are less likely to be completed because supervisory officials are apathetic toward monitoring ongoing projects away from the scrutiny of superiors; furthermore, the fate of rural projects is easily determined by the whims of local officials nourished by various ideological biases. Rural development in developing countries is also essentially compromised by a series of problems with implementation and completion of projects due to a "failure to understand the institutional dimensions of economic behavior at the village level . . . and ethnocentric analysis that is unable to connect intellectually with the world into which such projects are introduced" (Bromley and Cernea 1989: 27). Whether advocated by Western scholars and development agencies or by non-Western scholars, the common solution to accumulated rural problems seems to be the replacement of local strategies by Western technology (Bendix 1967; Pingali et al. 1987; Shiawoya 1986).

In the case of Nigeria, Kiwanuka (1986) captures the sentiment expressed by two camps: those saying that "to break out of the vicious circle of food shortages [Nigeria] must adopt a massive mechanization strategy and commercialize its agriculture (1986: 13), and the opposing view that the small-scale farmer must be the target in any effort to stimulate agricultural growth (McKone 1985; Dannhaeuser 1987; Kiwanuka 1986). The latter suggestion has been refined as "giving aid to the marginally rich who have demonstrated their ability to make good use of it" (MacDonald 1989: 99). None of these suggestions considers the environment or the adaptability of those cultural institutions in which the agriculture of the rich, the marginally rich, or the poor may thrive. Others perceive the need to enculturate non-Westerners to Western attitudes as prerequisites for development (e.g., Myrdal 1968). Consideration of modern technology is also central to the "humanistic" perspective of development popular with anthropologists who advocate adapting traditional modes to modern systems (MacDonald 1989; Escobar 1991). But the results of these approaches have had an unimpressive effect on the majority of Africa's inhabitants because of the divergence between the conceptions of projects and the local conditions. I do not attribute this failure exclusively to technology, because it is impossible to achieve economic advancement without modern technology. Indeed, local circumstances also embody important cultural elements and practices that discourage development initiatives by both external and local agents and constrain the realization of desirable changes in rural communities.

Social scientists have long since put aside the disdainful perceptions that the culture of indigenous peoples constrains their mobility toward economic and technological parity with the West—a view quite popular among earlier scholars of cultural change. Writing in the 1950s, Hoselitz asserted that traditional forms of agriculture, for example, slowed economic growth by resisting change (Ruttan 1988). And earlier writings of Malinowski and Radcliffe-Brown on South Africa attempted to fit indigenous people into a colonial frame in order to better control the "destructive" aspects of native cultures (Magubane and Faris 1985). These earlier views are often criticized as being Eurocentric and atomistic (Hoben 1982) because of their implication that existing institutions are constraints rather than the framework for development, or as advocating the replacement of traditional values which were seen "as retarding, with new, progressive, essentially Western ones" (Ramirez-Faria 1991: 117). While this criticism is valid, it is also idealistic in attempting to justify every indigenous cultural practice, particularly as many of us are searching for ways to effect significant changes in our lives. On the contrary, many of us who have to contend with the hardship in our villages see such criticisms as antagonistic to development, particularly when aspects of disciplines such as anthropology focus on the "preservation of indigenous people."

Today anthropologists increasingly give high priority to involving target communities in development projects from the planning stage up. The assumption is that chances of success are maximized by involving rural people in planning projects designed to benefit them (Finsterbusch and Van Wicklin 1989; Escobar 1991). Sound though this concept is, the actual project implementation process regularly discounts local social realities that are seen to conflict with the design of the policy. There is often a failure to consider the various cultural elements that may adversely affect development "from below," even when development agencies favor community-based development (Mackenzie 1992). Responsibility for articulating and incorporating the effects of local attitudes, practices, and expectations is placed on the "beneficiary" communities themselves. They must make the apparent "magnanimity" work.

For example, a road construction project was discontinued in one of our villages because villagers did not help government workers do the work. Local government workers who came to repair the road did not bring equipment for lifting the logs needed in the construction. They sent a message to the village head requesting that villagers be sent to help at the work site. The request was turned down for a number of reasons: it came at a time when we were clearing the land for planting;

farmers had assigned themselves into labor exchange groups with set schedules that the roadwork would have disrupted; the work site was four miles away from the village and the work required more time than we could offer at the time; and some of us could not leave our own farmwork to perform a task government workers were paid to do.

The project was thus terminated on the basis of the faulty perception that we lacked interest in it. Government workers ignored the cultural factors operating at the time that affected the response of villagers to this external development initiative. When the government considers the local culture at all, it is often hindsight, as the experts ponder the puzzle of the unsuccessful implementation of their plans. In this case, local labor was demanded without effectively coordinating the plan with the people involved. Members of the target communities may contribute labor effectively if local-level organizations are consulted by development agencies and empowered with local authority to oversee and sustain such projects.

Trivializing the Non-technological Factors in Underdevelopment

In other cases, "development economists while accepting the existence of non-economic factors in underdevelopment, have tended to put them on a 'lower order of variables,' or to use them only as discrete or ancillary aspects of the entire picture" (Ramirez-Faria 1991: 117). This lack of interest shows a selective interpretation of Leslie White's idea that technology is a pivotal aspect of cultural change to which other aspects of culture will adjust (1959); it considers such nontechnological features merely ideational and attitudinal components of culture, but White perceives them as necessary for implementing technological change (Hatch 1973).

These nontechnological factors have repeatedly been shown to influence the direction of rural people's response to change (Brosius 1988; Grayzel 1986). Grayzel reports that concepts of beauty and freedom among the FulBe of Mali figured significantly in their migration choices, their relationship with their neighbors, and their disinterest in a government project aimed at developing a grazing area for their cattle. Among the Biase, the ephemeral and symbolic elements of ideological and magical rituals, which are too often perceived as irrational, are as relevant to agricultural growth as the concrete elements of scientific input or "rational" production decisions. Farmers in some Nigerian communities still take their tools to priests for blessings before farmwork begins; and the festival of the new yam practiced in most of southern Nigeria expresses farmers' gratitude to the fertility deity for providing food (Anikpo 1986). However, such ideological adherence is not absolute. Anikpo observes that while traditional rituals and symbols have survived modernization, the authority structures they helped support earlier have been weakened. I will discuss later how the diminished belief of the Biase in the efficacy of magic and rituals as well as their irreverence for the supernatural have significantly affected their devotion to traditional ideologies. Modern technology is important but not all encompassing. Therefore focusing on technology as the main variable affecting response to change risks ignoring or trivializing the changed configuration of social structure and how this change affects rural economic growth.

I do not mean to say that if projects account for traditional practices they are necessarily successful. The Ujamaa strategy for rural development, for instance, was based on an assumed contract between equals at the village level. It stressed the need for communal living; a collectivity of villages was to form the unit of production and the nucleus of rural transformation. Unfortunately, the basic African unit did not live according to the principles of Ujamaa; traditional forms of cooperation were hierarchical rather than democratic (Lofchie 1978). The failure of the Ujamaa strategy thus resulted from a misinterpretation of current realities and a misunderstanding of the changes that have since evolved to challenge historical realities.

Internal cultural factors are important for understanding the Biase situation because isolated communities such as ours tend to explain and attempt to resolve their development problems on the basis of an indigenous worldview and traditional coping mechanisms.[1] Chances for development to be successful and meaningful to us depend on seeing how the requirements of technology fit within the context of our institutions, that is, the manner in which the concept of development fits local perspectives.

The problem scholars in development seem to face is one of defining terms in ways that are cross-culturally valid. Frequently the measuring standard is based on a Eurocentric ideal that contrasts development with stagnation (Ramirez-Faria 1991). Finding no alternative, Africans and Africanist scholars often perceive underdevelopment as the oppo-

1. See Hammond's study of the Mossi of Upper Volta (1959) and a study of the FulBe of Mali by Grayzel (1986).

site of development: the absence of features such as agricultural mechanization and the consolidated farmlands that are evident in more advanced economies (Kiwanuka 1986). The implicit assumption is that these indicators of the absence of development must be eradicated in order for progress to begin. Yet despite years of adherence to this doctrine, as well as the attempts and experiments that less developed countries themselves have made, the achievement of economic growth has remained elusive.

The lack of locally relevant development formulas has left the concept of development open to many interpretations. For agencies such as the World Bank and the International Monetary Fund (IMF), gross national product (GNP) weighs profoundly in a concept of development that aims to shrink the national balance of payment. While such models have constituted the guiding principle for economic development in many less-developed countries (LDC), they have had limited impact in much of Africa.

It is not only scholars in the LDCs who have problems with the concept of development. Western scholars face the same problem (Ramirez-Faria 1991). For example, I was once in a study group of American students in which everyone was assigned the task of coming up with a definition of development. We ended up listening to definitions of what development is not. A model for development that can be applied is still missing from the development package. I do not mean to say that a definition specific to a locality will solve the problem of development; but it is alarming that objectives are arbitrarily set for all cultural and ecological regions of the world on the basis of a term that remains ill defined but to which all the world seems to be expected to conform. Unfortunately, those for whom the development experts have set up this finish line (however ill defined) are only very faintly familiar with the track. This point is of particular concern to some LDC scholars, who stress that in "erecting Western societies as a model for the rest of the world to follow, accounts of modernization tend to assume that we know all about the West. Yet this assumption cannot be justified since we do not fully understand the relations between social structure, history, and symbolic-cultural meanings and values even in Western societies" (Fruzzetti and Ostor 1990: 22).

As this often ill-considered race for economic growth continues, the concept of development should be based on "indigenous understanding of the term, not on external definitions, for indigenous strategies vary from situation to situation" (Wright 1988: 382). Scholars have referred to development as improvement in the quality of life, although different cultures may define "quality of life" differently (Jack 1988). Just as the concept of development is defined variously by the experts, so do rural Africans variously define what is important to them—the presence of which will mark a community as "developed." There are rural communities in Nigeria, for example, where a person must make a journey by foot before arriving at the nearest road. Being able to make the journey by bicycle would be an improvement of the quality of life for such communities. More generally, it is hard to see any sub-Saharan African rural community that would not term itself "developed" if it experienced occupational diversification, the release of labor for nonfarming sectors of the economy, surplus food, ease in distributing and acquiring resources, good transport, such state-provided assistance as good drinking water, and health delivery services. Western scholars might add other factors, such as political stability, at least eleven years of education for 30 percent of the population, higher per capita income, and improved levels of nutrition.

Despite local variations in the perception of development, our desire for change is generally articulated in terms of relieving stress rather than eliminating poverty, of regulating today's challenges rather than conquering tomorrow's uncertainties, and of integrating current practices and behaviors into traditional social institutions rather than preserving them as treasured guardians of once-effective social institutions. Whenever we make supplications to our deities and spirits for wish fulfillment we reinforce these simple premises.

When we visit friends, community leaders, or elders, we often pour a libation (plate 27) and the recurring prayer to ancestral spirits is usually that they should make today better than yesterday and tomorrow better than today. This elementary supplication is at the core of what we mean when we think of development, although there is necessarily a value premise to its application. In Biase we pray for a good harvest, abundant fish, and the ability to transport our surplus to the market. The city corporate executive, using the same general formula, might define development in terms of a few dollars more than last year's figures, or a faster computer.

Rural development involves a modification of both material and nonmaterial factors. It occurs when a community initiates an adjustment of the ideational and attitudinal components of its culture so that enhancing factors are activated for improving an existing state of affairs. First there must be a cognitive disposition of the target community to modify its affairs and mobilize adjustive elements by maximally utilizing men's and women's labor, ensuring significant participation of men and women in the economy, and reassessing traditional rights and privileges.

Sometimes, however, existing local regulatory organizations are handicapped by traditional practices and behaviors that conflict with newly evolved elements, thus reducing their effectiveness. The men's Abu society in Biase, for example, performed our judicatory functions until the end of Nigeria's 1967–1970 civil war. After the war its elders were constantly threatened by Biase youths freshly returned from the war. Some of the youths who were discharged from military service accused some of the elders of using magic to get them discharged from the military at the end of the war. Abu became helpless under this threat from young radicals and was consistently derided by other members of the community for its inability to enforce the traditional regulatory rules that earlier, communal members had rarely deviated from. As a result, Abu has become powerless against the new demands of gender and age and a whole range of cultural factors which collectively challenge its legitimacy.

It would be wise to pay attention to these limitations and to account for them in the implementation of projects because when realities on the ground and the perceptions of rural people conflict with the means of the realization of goals enforcing a standard package design is unlikely to work well (Green 1989). Indeed, as Ruttan asserts, "when peasants refuse to adopt the practices recommended by agronomists and economists, it may be the experts rather than the peasants who are wrong" (1988: S256). The bounds of indigenous life are exposed to all forms of external social and economic invasion. In order for us to function more effectively within this changing environment and confront increasing modernization, it is important that we become better-informed participants. We could begin with a closer focus on the adaptable nontechnological cultural elements that influence the economic processes of our communities, and by deemphasizing paternalistic interpretations that only contribute to rural neglect. Factors such as the leisure time of rural men, the absence of men from rural markets, the low involvement of rural people in the national marketing network, and beliefs and ideologies that affect production need to be understood in historical context and as they function now. On another level, rural development requires remedying those constraints to development that are beyond the control of the target community, such as economic and political neglect by the state, lack of access to institutions and services necessary to foster local economic growth, and disdain for the nontechnological dimensions of development.

Development Policy in Sub-Saharan Africa

The irony of development policy over the years is that a significant sector of the population has been required to reduce its hope of ever experiencing the results of development because tools and accessories have had to be concentrated in the hands of a few "experts," who often need a few decades to hone their proficiency. Presumably such postponed expectations will result in mass benefit in the long term. This policy seems to fit the Biase ideology of rural economic rationality because we often toil in the hope that we will build better economic conditions for our children. In earlier times, our various lineages were often ready to contribute money for the education of children that were expected to be the instruments of communal change, and lineage members would forgo the short-term advantages of alternative uses of their money.[2] The impetus for this self-sacrifice was the hope of grooming persons expected to assume government positions and subsequently add our villages to the priority list for government attention. Often there is pride in the knowledge that a particular government official is a "son of the soil," although his or her community may never benefit from government policies.

In this way, sub-Saharan African governments have often successfully tapped into the ideology of acquiescence among rural people, which is forgiving of the general absence of state-provided amenities. Years later, rural people such as the Biase who have not experienced the advantages they expected from their past human investments tend to be less enthusiastic about collectively sponsoring individual successes. The paradox of rural development in much of Nigeria is that while dismal rural poverty usually justifies research and aid grants, only a handful of urban Africans benefit from the grants. For as Haugerud (1989) reminds us, agricultural projects in countries such as Nigeria are usually directed toward the benefit of large-scale farmers at the expense of smallholders, and extension services mostly favor producers of cash crops. It is a dilemma that applied anthropologists and development practitioners continue to wrestle with.

Although anthropologists are familiar with interference from government officials and with having their recommendations rejected by

2. Bates (1990) reports that parents in the Kasumpa village of Zambia spent a lot on the education of their children in the expectation that they will later receive financial returns in the form of remittances from the town on their investments.

governments because of "administrative prejudices and preference for familiar policy recommendations" (Partridge and Eddy 1978: 25; see Mason 1990),[3] their interest in rural development studies remains undiminished. Hoben says anthropologists have been significant in clarifying and challenging erroneous assumptions made by development planners. This advocacy stance may have created the perception that the anthropologist had a "place in the village and that his only contribution to development was to serve as interpreter" (1982: 354). But present-day anthropology is more involved than this statement seems to convey.

The anthropological perspective on development, as summarized by Hoben in his review article (1982), shows that anthropology has continued to wrestle with both earlier perceptions of indigenous people as tradition bound and a later perspective of the "rational peasant." Numerous studies show that indigenous people are no more tradition bound than Westerners, and that their productive systems are indeed adequately adjusted to their circumstances. Consequently emphasis has shifted from such unpopular concepts as the risk aversion of traditional farmers, which is usually attributed to the proven adequacy of existing indigenous methods and the farmer's poverty (Rogers 1969; Mellor 1969; Barry et al. 1959; Schultz 1964). There is now growing interest in models that show farmers as likely to resist modern technology that is unsuited to local conditions and that increases dependency, but as ready to accept reasonable changes (Belloncle 1973; Wade 1974; Knight 1974). A complementary perspective says that farmers make rational economic decisions and will adopt new practices given sufficient incentives (Johnson 1972), or that they consider innovations on the basis of their relevance to their circumstances (Hoben 1984; MacDonald 1989).

From the late 1950s perspective that tended to blame target communities for failure of development initiatives, the focus has shifted to paying greater attention to the needs of communities : a "project begins to take shape when, upon evaluation of the cultural characteristics

3. Mason reports the bureaucratic constraints he encountered while working as part of an interdisciplinary team designing a national settlement plan for Libya. "I had frankly underestimated the political abilities of certain members of the planning profession, especially the project manager, who brought with him over four decades of experience in town-council bureaucratic machinations and was a master of bureaucratic intrigue" (1990: 169). The project manager sent cables in the name of the country's deputy minister to the sponsoring agency in New York without the minister's knowledge, "issuing formal instructions on critical policy and management issues" (1990: 169).

of the group, the possibilities of alternatives are determined. The elaboration of a project seeks to maximize the realization of community interest" (Wright 1988: 370). When a community's needs are understood, efforts to satisfy those needs are more result oriented and less cumbersome as target communities identify with projected goals and strive for success.

A remarkable example of this occurred while I was in the field. Because of a new government policy that aimed to alleviate harsh rural conditions, a group of government workers was commuting daily to one of our villages to execute a borehole water project. On the third day of work, the drilling tool the group was using broke and a piece was lodged below the twenty-foot hole already drilled. Work on the project stopped for three months because the group was unable to dislodge the broken piece of equipment. The group shuttled between the village and the city in search of experts to help with the problem. Meanwhile, fascinated villagers, who always gathered around the project site, maintained their curious vigil. Occasionally, the workmen would request one of us to perform such menial errands as buying them cigarettes from the local stores. Since work had been virtually stopped, the workmen spent more time talking with us and meeting with some of us for drinks. It was during one such meeting that the problem of the broken drill became clear to some of the villagers. A carpenter in the village suggested to the skeptical government workers that he might be able to dislodge the broken piece of equipment and convinced them to lower him into the hole with some of his tools. He secured the broken equipment with wood and nails, attached a rope to it, and asked the villagers to pull. After three attempts, the four-foot-long piece was dislodged, to loud applause. Within minutes, the entire area was swarming with villagers who had heard the news and came to congratulate the carpenter and the other villagers who had helped to solve the problem. Thus, when the involvement of villagers is not limited to the performance of menial tasks, communal work groups are enthusiastic and can be scheduled into the organization of work as part of a planned cultural adjustment.

In present-day Nigeria, many rural communities, stimulated by a few wealthy migrants to cities, are embarking on various projects aimed at imitating city life. There are projects for rural electrification, pipe-borne water supply systems, health care centers, highways with concrete bridges, and just about everything governments believe rural people do not need. But Nigeria's poor economy is making it difficult for many rural communities to manage such locally initiated projects, as well as limiting the ability of emigrants to send home outside goods or to send remittances to support the economic effort of relatives back home. Poorer communities then feel the movement toward modernity only too feebly. Despite the fervent cries of community elites, who from comfortable city abodes bemoan the changes in our culture and plead for the preservation of our cultural heritage, there are hardly any communities content to remain at traditional levels when other similar communities are perceived as benefiting from modernization.

Clearly, the Biase perspective on development reflects our desire for a change from difficult conditions. Development is described in various southern Nigerian languages as "going forward," "dawn," and "getting off the ground." These terms seem merely descriptive of the economic circumstances among different ethnic groups and describe aspirations toward remedying a general lack of occupational diversity, general immobility of labor, political instability, little or no education for 70 percent of the population, problems of distributing and acquiring resources, poor roads, and the absence of either state or communally provided social amenities.

Presumably, the standard development criteria, particularly the tradeoffs mostly favored by the governments of developing countries, have disadvantages which are supposed to be temporary and self-correcting: eventually the benefits of development will trickle down to the poor. Unfortunately, little ever trickles down to the poor because while "leaders of underdeveloped countries are aware of the impact of development, few are able to separate the economic from social aspects, and by emphasizing the economic, incur a damaging cost to the people" (Fruzzetti and Ostor 1990: 17). For example, in Nigeria's Bakolori irrigation project, the "calculation of future economic benefits in the design depended on the attainment of dramatically increased crop yields" (Adams 1988: 319). There was an overriding faith in the economic returns from a sugar estate that was expected to benefit from the program, but the economic returns were never realized, and conflicts between the government, the project contractors, and the villages affected by the project about appropriate procedures created hardship for the villagers. Donnelly notes the same problem in Brazil. He reports that by 1979, Brazil was able to raise its GNP to one of the highest in the world, but it was inattentive to short-run poverty problems as reflected in increased levels of economic and material deprivation. Following conventional wisdom, Brazil had grown to a point where "benefits should be not merely trickling, but cascading down to the poor" (Donnelly 1984: 260). They did not. Also, as Nigeria's "oil production reached two million barrels a day in 1973 and oil prices quadrupled during 1973 and 1974, Nigeria entered a breathtaking economic boom" (Diamond 1988: 45). This did little for farmers because the national development plan launched in this period had no specific programs for the agricultural sector (Anyatonwu 1986). This is a problem rural communities continually face.

The favorite focus of development and development theory is usually the large political culture. This has tended to constrain understanding or formulation of policies relevant to rural people and their modest development needs. In many developing countries, just as in Nigeria, national development policies concentrate on building up urban areas and providing them with amenities—a bias attributed to the perception of rural areas as reservoirs of resources for urban development.

Some scholars have expressed doubt about the ability of international agencies to advise the developing world on its agricultural development problems because their officers have no direct practical experience of the farming systems of rural cultivators (MacDonald 1989). MacDonald notes that international development agencies hurriedly hand out loans in order to avoid large interest payments. "The fact that few projects that are implemented ever achieve the projected rates of return does not seem to matter: as long as the study shows that a good return is feasible then the agency is happy" (1989: 96). He discusses organizations such as the FAO to show how such agencies depend on computers to solve the problems of developing countries:

> For one project, it was necessary to estimate the present production of some twenty thousand subsistence farmers, but no reliable estimates for the areas of crops grown and the yields were available. In fact more than twenty crops, all mixed together within the same parcel of land, were grown by the individual farmers on their holdings, some consisting of only a few square yards of a crop such as herbs or spinach. To resolve the position five major crops were used to represent the probable crop production, which could then be translated into a money value."

This was then used to calculate the benefit of the new farming system to the farmer. As the system was not controlled for crop varieties by individual farmers, the planners ended up with several thousand tons of cassava, which rendered the entire exercise worthless (MacDonald 1989: 98). The implication is that such externally packaged projects only benefit the development agencies, because they breed a proliferation of experts who are independent and in competition with each other for academic and official rewards (Okoli 1989).

Faith in the standard criteria for development with the related misreading of facts on the ground is the doctrine not only of foreign development "experts" but also of African "experts." Commenting on the slow pace of Nigeria's agricultural development, Anyatonwu recommends that "the Federal Government agricultural policies and contributions for the 80s must deemphasize the public sector and small farmer schemes in preference for large-scale commercial agriculture that could effectively solve the country's agrarian (food) problems" (1986: 72). According to him, mechanized farmers, rather than small-time farmers, should be guaranteed financing, since they are more profit oriented than small-time farmers (1986: 72). Others blame the farmer for Africa's food crisis because he "is a totally frustrated and disillusioned man" who cannot be relied on to feed the nation (Nwankwo 1981, cited in Kiwanuka 1986: 14).

Such misunderstanding sanctions the obsolete perspective, referred to earlier, which sees us as conservative and tradition bound. Anthropologists now mostly favor the notion that traditional institutions are often adaptive to a changing socioeconomic environment. Small-time farmers "accept the challenges of innovations that improve their livelihood" (Fruzzetti and Ostor 1990: 18) and should be made the primary locus of change (Pitt 1976; Low 1986; Longhurst 1988). Johnson (1972) attributes the idea that farmers are conservative to the false belief that rural farmers rarely deviate from traditional norms and patterns; indeed, such an interpretation has been an important reason for the neglect of rural communities by African development planners.

The alternative model of development which seeks to use indigenous knowledge as a referent is sensitive to the need both to preserve the environment and to involve rural peoples from the ground up. This means that they must be "trained to build appropriate adaptations into their existing environmental knowledge and skills" (Blaikie 1989). Of course there is still the question of which community members to focus on if development is to tap indigenous skills. And projects are quite often geared toward what is needed to oil the wheels of government rather than what the community really needs.

In addition, project designs that aim to alleviate rural poverty sometimes fail to account for the possible responses of existing rural institutions. An example (Bromley and Cernea 1989) is the First Livestock Development Project in Botswana, in which the government introduced land zoning arrangements that were at variance with existing agricultural practices. Communal lands were used to establish large ranches, with the assurance that adequate "communal" land would remain available for the needs of those not involved in the ranches. This expectation was not achieved because Land Boards set up by the government to replace local chiefs in assigning grazing rights were unable to regulate the uncontrolled access of commercial ranchers to the communal lands. Consequently the designs of such development projects conflicts with local realities even before their implementation, because no local organizations are in place to overcome difficulties or sustain initial efforts after projects are concluded.

Conclusion

Our former coping mechanisms have lost their effectiveness; living conditions in our villages have deteriorated without external support to complement indigenous knowledge and institutions. Every community needs help to respond to a changed environment that poses survival problems unresponsive to indigenous coping strategies. Often anthropologists are more familiar with these circumstances and their consequences for underdevelopment than are other scientists. But their avoidance of advocacy does little to alter the perception that the discipline's only contribution to development is preserving indigenous cultures as "objects of anthropological possession and the testing of theories" (Wright 1988: 369)

Non-Western peoples have gone through enormous transformations in the past several decades. Changes have occurred in both social and cultural institutions. My argument is not that previous conditions were better than today's but that previous conditions presented fewer challenges, and that people were able to cope more effectively.

Our people do not agree that we will be better served if our institutions are left intact. We seek and aspire to replicate the attractive ways of life of urban dwellers. During a visit of the Cross River State governor to our local government headquarters, the most persistent refrain in the addresses of welcome presented by villagers was that the government should provide good roads, health care facilities, a pipe-borne water supply, and electricity. We need to better understand these changes in rural structure and the desire for modernity which they reflect in order to understand how traditional organizational arrangements guide the responses of rural peoples, such as the Biase, to the pressures to modernize.

TWO

The Biase of Southeastern Nigeria

Among My People

> Fieldwork in one's own culture has the advantage of allowing one to develop more insight into the culture because of familiarity with it and to arrive at abstractions from the native's point of view. One does not have to learn another language or understand a different way of life, and one may have little difficulty in developing rapport . . . Subjectivity in studying one's own society is inescapable, especially if the fieldwork deals with issues that are emotionally charged.
>
> Kim Choong Soon, "The Role of the Non-Western Anthropologist Reconsidered: Illusion versus Reality," *Current Anthropology* 31 (2):196–201 [1990].

Reading Kim's article on doing fieldwork in his native Korea made me feel as though he were looking over my shoulders as I did fieldwork among the Biase (fig. 2), the Nigerian minority ethnic group to which I belong. Participating in the various activities of the Biase was satisfying not only as a research approach but also as a means of reimmersion in a culture I had left behind a few years back. Since I had been away from the Biase for about eight years, my impressions of the group tended to be influenced by conceptions derived from written works on other rural people. Consequently I constructed assumptions about the Biase from my familiarity with both Biase culture and the current literature. Significant changes had occurred in Biase since I left. In particular, major landslides have contracted some of the coastal villages, making them look smaller than I remember. The only thing that seemed unchanged was the destitution all over Biase.

The thought that often came to my mind when I remembered Biase was its isolation. People were generally without any of the amenities commonly found in the cities. Those visiting from the cities often seemed more prosperous than those in the villages: their city clothes, their fancy cigarette brands, their taste for city beer rather than the locally brewed palm wine, a little gift of money here and there, and their

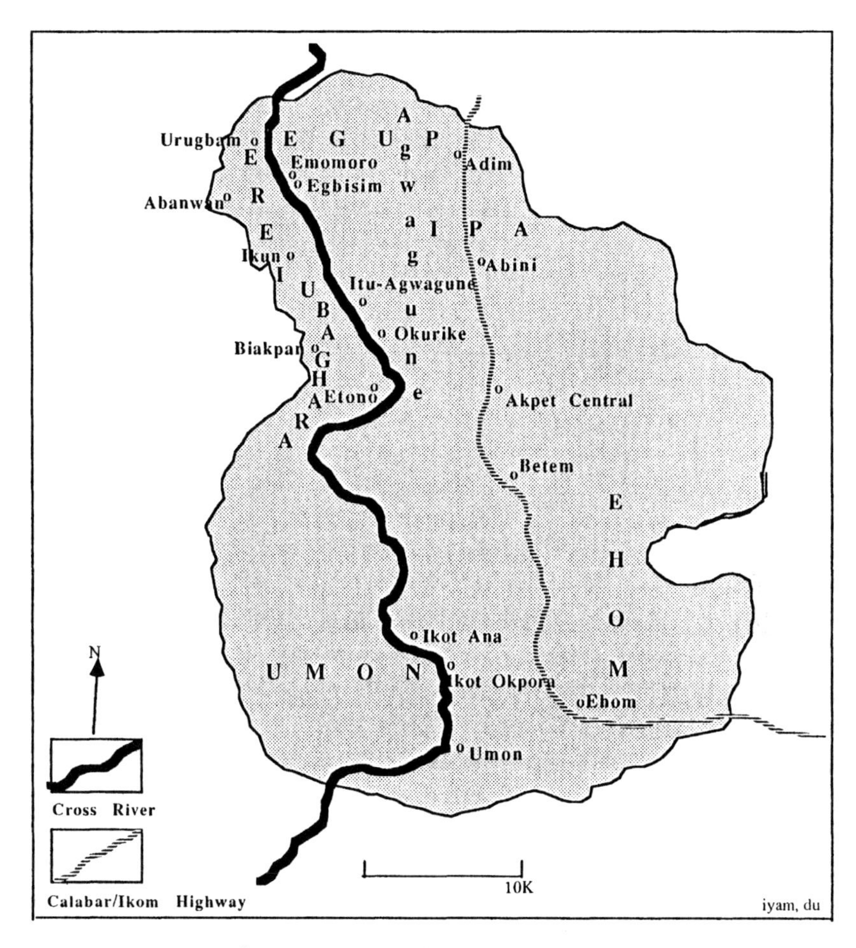

FIGURE 2. The Biase Group, Showing the Cross River and the Calabar-Ikom Highway

general air of superiority all tended to place them above village folk. It was therefore not unexpected that my reentry into Biase society was locally billed as much more than anthropological fieldwork.

My arrival at my primary field site resulted in many activities meant to formally welcome me back to my Agwagune community. Over a period of six weeks there was a series of ceremonies by friends, age set members, men's and women's organizations, and my patrilineage and matrilineage. I strongly objected because I knew people were putting themselves out to sponsor these events. It also cut into the time available for my research. By the end of the first week, I was able to convince some of my well-wishers to hold off their welcomes, including a joint ceremony planned by all Agwagune villages. I told them the ceremonies would be inappropriate since I was only visiting Nigeria and not returning finally. I also said that the ceremonies would be more gratifying if my wife and children (still in the United States) were present.

During my first two weeks in the field, I traveled through the five Biase communities in order to get an idea of the general characteristics of the villages and to develop an index for measuring economic growth. I noted the presence or absence of social amenities , the political strength for negotiating with the larger polity, the technology for addressing food shortages and ecological problems, the availability and accessibility of such government provisions as dispensaries, a clean water supply, an adequate educational infrastructure, electricity, and the police.

As is common in rural Nigeria, these economic growth indices were generally absent. The exception was Akpet, a village of about three thousand inhabitants that operated a government-run health facility. Major differences between Biase villages and other rural communities include distance from urban centers,[1] minimal economic cooperation and little transportation to outside communities, and flooding.

Early in my research, many people expressed so much skepticism that it almost marred my census taking. They believed I was just another government official sent to assess their needs. The previous year our villages had suffered severe flooding that had resulted in a great loss of property, homes, and farmland. A community representative had assessed the damage and told flood victims that the government said that each lineage should contribute a three hundred naira[2] deductible sum toward the cost of applying for government restitution. Until the time of my arrival nothing had been heard from the government. Some therefore expressed disgust that they had to face yet another futile session of questioning regarding their losses.

However, I conducted a census of my primary site during the second week. This gave me an opportunity to meet people, renew acquaintances, and assess changes people and households had experienced in the past years. The census took longer to complete than I had expected because many people simply wanted to talk and ask me ques-

1. The nearest city, Calabar is about 100 kilometers from some villages.

2. The Naira is the Nigerian monetary unit. One Naira = 100 kobo (1990 dollar exchange rate fluctuated between U.S.$1 to 10 Naira. The Naira fell to about 35 Naira to U.S.$1 by the end of 1993.

tions about the "white person's home." In this role reversal, I spent long hours answering questions from my people about the white person's home and editing my tales just enough to de-emphasize the dangers of living in Los Angeles.

Because of unanticipated problems, it took two weeks longer than I had expected to take the census. I had a female, Egwa Ogozi, and a male field assistant, Ikenga Abba, working with me on enumeration. We divided each of the two communities into three segments; everyone had a segment to enumerate. The sample frame for Odumugom included a total of 269 households, and there were 437 households for Abini. My census included questions on labor history, number of years of education, off-farm occupation, number of farms owned, crops cultivated, number of adults and children living in a household, household composition, and migration history. During the first month of my study I also made an inventory of significant materials in each household, noting the presence of such items as radios, television sets, and European-style furniture. I conducted a second census in October about the time of the new yam festival to see how the visitors and returning migrants who flooded the villages for the ceremony might have affected the economic landscape. From my first census, I drew a random sample of 60 households (30 each from Abini and Agwagune) which included 224 persons (92 women and 132 men).

During the enumeration, my field assistants kept reporting that some people were not cooperating and were giving all kinds of excuses to keep them away. I later visited the noncooperators to find out what was going on. They generally expressed displeasure that I had sent my field assistants to interview them instead of visiting them myself. Subsequently, I scheduled time for visiting just about everyone in the community during the period of the census. This resulted in visits to about 600 households over a period of four weeks. My time did not go unrewarded since it afforded me an opportunity to look around homes and get a sense of household possessions.

A more significant dilemma was that people were on their farms most of the day and came back late at night. With the poor natural light, it was difficult to work most evenings, although this was the best time for village people. This was also the time women prepared dinner, while men congregated around patrilineage centers drinking palm wine. I constantly turned down invitations to stay for dinner or share a cup of palm wine. Some people perceived this as aloofness resulting from my stay in the white person's home, which did not help me to build trust at the beginning of my research.

It was difficult for my Biase friends to understand that I was in the community to study aspects of our culture because they assumed I already knew it well. In the same way, most people considered themselves cultural experts. After a random sampling of the study households from each of the two communities, I faced the predicament of explaining to the other households why I was not going to use them much in my study. People whom I had not visited often reminded me as I walked along the street that they were still expecting me. Rather than offend more people, I decided to visit people outside of my sample and use them to triangulate information from my key informants. Such triangulation became important as my research progressed, making it easier to determine a community pattern in the conduct of daily affairs while at the same time fulfilling my traditional courtesies as a member of the Biase.

Unfortunately, some of the informants I decided to use had displeased community members in various ways. One of them was a forty-five-year-old man who was not only always drunk but bought everything on credit and never paid his debts. Another was a forty-eight-year-old handicapped man who was also believed to be a thief. The townspeople were shocked to see me visiting these people in their homes and associating with them regularly. People complained I was lowering my status by fraternizing with these people and sent friends to advise me to find more respectable company. As Kim notes in his study, when he fit the Asian stereotype, his people treated him as a foreign guest, and when he failed to do so, his company was less welcome (1990: 198). This was the most difficult experience I had in the field, partly because of the communal disappointment I perceived and partly because I felt an obligation to my culture. People did not usually see me as studying the culture but as participating in it; that level of participation was expected to be complementary with the status of one who had "fraternized with the white man for many years." I was regularly scolded by people in my paternal and maternal lineages who worried that I was a local star who let his brilliance be dulled by the community's dark clouds. I created a better understanding when I explained the nature of anthropology to the people who had expressed concern, thereby justifying my apparent relationship with the socially unloved.

Another predicament resulted when I bought a small electric

power generator for use with my laptop computer. This prompted people to request that I run electric cables to their homes. Although they did not seem to understand, I explained that the generator was too small to supply more power than I needed to carry out my work. There were often adults and children blocking my doorway to satisfy their curiosity about the wonders of the computer. Other people thought I needed their company and felt obliged to spend most of their time with me. I often felt compelled to make excuses that would persuade my visitors to leave; they never seemed to understand that I needed time to do my writing.

Visitors to Nigerian rural communities are often encouraged to call on the village chief first. I was told of a continuing chieftaincy dispute in one of the two communities I was studying. One of the chiefs had formal recognition from the state government but was rejected by over 80 percent of the people, while the leader recognized by the people did not have the support of the government. On my first visit to the community I was directed to the government-recognized chief. Following his advice, I waited two days for him to announce my arrival in the community before I started my research. I used the waiting period to look around the villages, introduce myself to people I did not know, and ask informal questions. Particularly helpful at this time were the meeting spots of the men, the patrilineage centers. In these informal arenas, I got more insight into political issues in the community. Much of the talk always had to do with the chieftaincy dispute. I got a sense of how bitter people were about the government-appointed chief. I was informed that associating with him would get me little cooperation from people. I later sought an audience with the popular chief, who immediately summoned his subordinate chiefs and introduced me. The town crier then went round the villages announcing my presence and telling the people I would be around to speak with them. Only after establishing this relationship was I able to conduct a census in that particular community.

Kim (1990) observes that the insider anthropologist suffers from his fellows' expectations of cultural conformity. Being a Biase presented a unique problem regarding the information people were willing to give, just as my research assistants and I felt uncomfortable about some of the questions we asked. People overreported or underreported such items as the number of farms owned, income from the previous year, number of children, and how many meals they had per day. I associated this misinformation with the reluctance of locals to have such sensitive details as personal income become common knowledge in the community, and to their holding back information they believed might affect their taxes. I had to reassure them that nothing detrimental would result from their answering my questions.

People were generally excited to hear that I was going to write a book about the Biase. Some seemed to believe that since I was doing the writing in the white person's home, the government would attach more importance to the book than if I wrote it in Nigeria. This meant the government would immediately provide such amenities as roads, bridges, and markets. I had the unpleasant task of informing them that my research might not produce such dramatic results. Once people understood that my study was not going to fulfill all their expectations for their communities, my research proceeded with little difficulty and much less interest.

Initially I spent much time in informal interactions, asking questions even about domains in which I felt I was fairly competent. The discussions often became intense as people tried to have their views recorded. They often expected me to ask them questions that would elaborate a certain point of view; some would insist that I write down their statements. These sessions produced considerable corroborated data and also gave me an opportunity to cross-check information from key informants.

Journeying between the two communities I worked in was a bit dangerous. To go from the primary community of Agwagune to the second community, Abini, I had to walk a 9¼-kilometer distance along a secluded road past the town of Adim, which was at war with my community, the Agwagune. Fearing that I might be attacked and killed by the Adim, the Agwagune elders advised me to go through a war ritual meant to protect me from machete cuts and bullet wounds. For seven days and nights, members of the Efa war shrine cut marks and rubbed dark powder on my hands, feet, and chest, brewed a mixture which I drank, poured burning liquid in my eyes, and cautioned me not to bathe for seven days. At the end of all this, the priests gave me an armband and a necklace designed to dull the machete and turn the bullets into powder. Luckily I did not have to test the efficacy of this ritual. However, it gave me a strong sense of security every time I passed through Adim and revealed to me how little all my years in the United States have affected my worldview.

I attended meetings of my matrilineage and my patrilineage where complaints were made and settled about familial and interlineal disputes, farmland distribution and encroachment on farmland, and work organization or the lack of it. I accompanied various patrilineages to mark out farmlands and to prepare the bushes for farming. I participated in birth and death rituals, rites of passage, and trials by ordeal to determine guilt or innocence, I witnessed cases and judgments about theft, marital neglect and infidelity, accusations of witchcraft, and war. I suffered a debilitating malaria and once again experienced treatment by native medicine men, my mother, and my friends. I became disillusioned about the many factors contributing to the poverty of Biase and agonized over the objectivity of my research.

Subjectivity was my most recurrent concern whenever I discussed my research with colleagues. I was aware that it would be difficult for me to do an unbiased study of my culture, but I did not know what direction my bias would take. Because of my focus on the cultural sources of underdevelopment and the assumptions I had constructed about cultures such as mine from the current literature, I worried that I would seem to be insensitive to the behavior and practices of my own culture. However, during my fieldwork I was painfully relieved to observe that some of my assumptions about the Biase were wrong. Cultural practices I had assumed were still fundamental to Biase society had been discontinued, reconfigured, or replaced. For twelve months I observed and participated in cultural practices that the Biase were still formulating and trying to understand. I tried to comprehend the poverty and destitution around me and to accept the fact that at the end of my study I would leave Biase as nothing "more than a predator consuming data" (Helge Kleivan, cited in Wright 1988: 365).

Background

Biase (fig. 2), comprises five subgroups, Ehom, Egup-Ipa, Erei, Umon, and Ubaghara, that lie in the poorly drained plains of the Cross River basin, situated within mangrove swamp forests that fringe the banks of the river (plate 3; fig. 1). Earlier references incorrectly identified the Biase as Akunakuna (Forde 1939, 1956, 1965; Talbot 1960; Harris 1965), probably because of the influence of a once-powerful Biase subgroup, the Agwagune. The Nigerian government's population bulletin lists the Biase population at about one hundred thousand people (table 1) scattered over sixty villages, with the greatest concentration in Egup-Ipa: over forty thousand. Some of the villages are marked by political rather than geographical boundaries because of the strategic necessity to defend specific communities. For example, Abini (fig. 3), a community in Egup-Ipa, comprises the villages of Edodono, Emomoro, and Afifia,

TABLE 1 Biase Population

	1963	1983	1984	1985
Subgroup				
Ehom	2,917	4,778	4,899	5,021
Egup-Ipa	20,639	33,807	34,664	35,532
Erei	5,283	8,654	8,873	9,095
Ubaghara	9,297	15,228	15,614	16,005
Umon	11,308	18,522	18,993	19,465
Total	49,444	80,989	83,043	85,118
	Egup-Ipa Population (Including the Agwagune Subgroup)			
Villages				
Adim	5,884	9,638	9,883	10,130
Abini	3,086	5,055	5,183	5,313
Akpet 1	1,459	2,390	2,451	2,512
Akpet Central	2,333	3,822	3,918	4,016
	Agwagune Subgroup			
Abamba	126	206	212	217
Abapia	300	491	504	516
Abaribara	509	834	855	876
Abredang	579	948	972	997
Abrijang	876	1,435	1,471	1,508
Egbisim	985	1,613	1,654	1,696
Emomoro	879	1,440	1,476	1,513
Itu-Agwagune	210	344	353	362
Okurike	1,517	2,485	2,548	2,612
Ugbem	1,161	1,902	1,950	1,999
Ijom	735	1,204	1,234	1,265
Total Agwagune	7,877	12,902	13,229	13,561
Total Egup-Ipa	20,639	33,807	34,664	35,532

Source: Ministry of Finance and Economic Planning, *Cross River and Akwa Ibom State Population Bulletin 1963, 1983–1990,* Statistics Division, Calabar, June 1985.

whose boundaries are marked only by bush trails, buildings, or ancient artifacts that have lost their original use; Agwagune villages such as Emomoro, Egbisim, Itu-Agwagune, Okurike, and Ugbem also have minimal geographical boundaries. Each village is politically autonomous and functions around a loose centralized authority represented by the Onun[3] who is appointed by the Inun of the patriclan. With a population of about four thousand Odumugom, comprising the villages of

3. *Onun* (plural *Inun*) is the title that designates heads of lineages (*Onun Ima*), villages (*Onun Emomoro*), groups (*Onun Agwagune*), and social groups (*Onun Abu, Onun Aneba, Onun Ebrambi, Onun Guinea*).

TABLE 1 Biase Population (continued)

1986	1987	1988	1989	1990
5,147	5,276	5,408	5,543	5,681
36,421	37,332	38,263	39,219	40,200
9,322	9,556	9,793	10,039	10,290
16,406	16,816	17,236	17,667	18,109
19,953	20,454	20,966	21,490	22,026
87,249	89,434	91,666	93,958	96,306
		Egup-Ipa Population (Including the Agwagune Subgroup)		
10,383	10,643	10,909	11,181	11,461
5,446	5,582	5,721	5,864	6,011
2,575	2,639	2,705	2,773	2,842
4,117	4,220	4,325	4,433	4,544
		Agwagune Subgroup		
222	228	234	239	245
529	543	556	570	584
898	921	944	967	991
1,022	1,047	1,073	1,100	1,128
1,546	1,584	1,624	1,665	1,706
1,738	1,782	1,826	1,872	1,919
1,551	1,590	1,630	1,670	1,712
371	380	389	399	409
2,677	2,744	2,812	2,883	2,955
2,049	2,100	2,152	2,206	2,261
1,297	1,329	1,363	1,397	1,432
13,900	14,248	14,603	14,968	15,342
36,421	37,332	38,263	39,219	40,200

Emomoro and Egbisim (fig. 4), has fourteen Inun while Abini has four village Inun to govern its population of about ten thousand spread over three villages, Edodono, Emomoro, and Afifia.

Murdock (1959) groups Biase (which he refers to as Akunakuna) under the Bantoid linguistic cluster of the Nigritic stock. The Agwagune language is spoken in five Biase villages, while other villages speak mutually understandable dialect variants.

Although literature is scant on the Biase (Attoe 1990; Ubi 1986), they are reported to share a few cultural and ecological similarities with their neighbors, the Efik (Nair 1972), the Afikpo (Ottenberg 1968), the Mbembe (Harris 1965), and the Yako (Forde 1965). These accounts report a sense of cultural cohesion and political strength that was so effec-

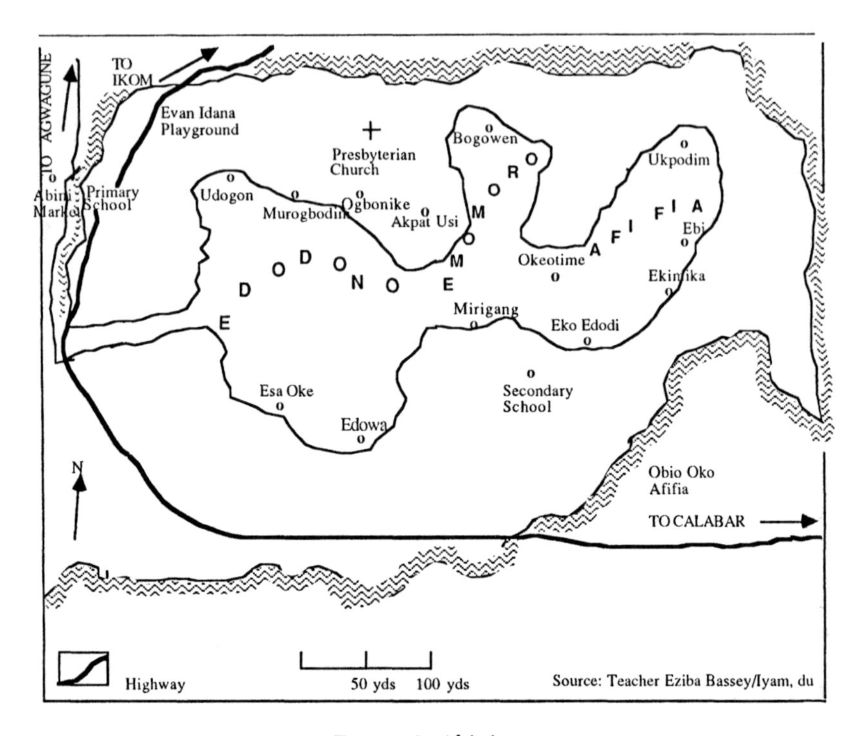

FIGURE 3. Abini

tive that the British colonialists had to send an expeditionary troop to the Agwagune to subdue and bring them under colonial rule (Nair 1972). In February 1896, "a force of 130 men under Captain Roupel marched from Calabar via Uwett to Okurike and thence to Emuramura Ekpesim, the principal town of the Akunakuna" (Talbot 1960: 229).

The Biase are hemmed in by the Cross river, which empties into the Atlantic Ocean, and a dense rain forest which offers little invitation to outsiders. Fishing and farming are practiced within this environment which provides subsistence to a population that reached forty thousand only by the beginning of the 1960s (Cross River and Akwa Ibom State Population Bulletin 1983–90).

ECOLOGY

Viewed from above, the joint Agwagune villages of Emomoro and Egbisim commonly referred to as Odumugom (Old town) look as though they had accidentally dropped from someone's hands and had broken to pieces, with the largest piece surviving in the shape of a new moon. This

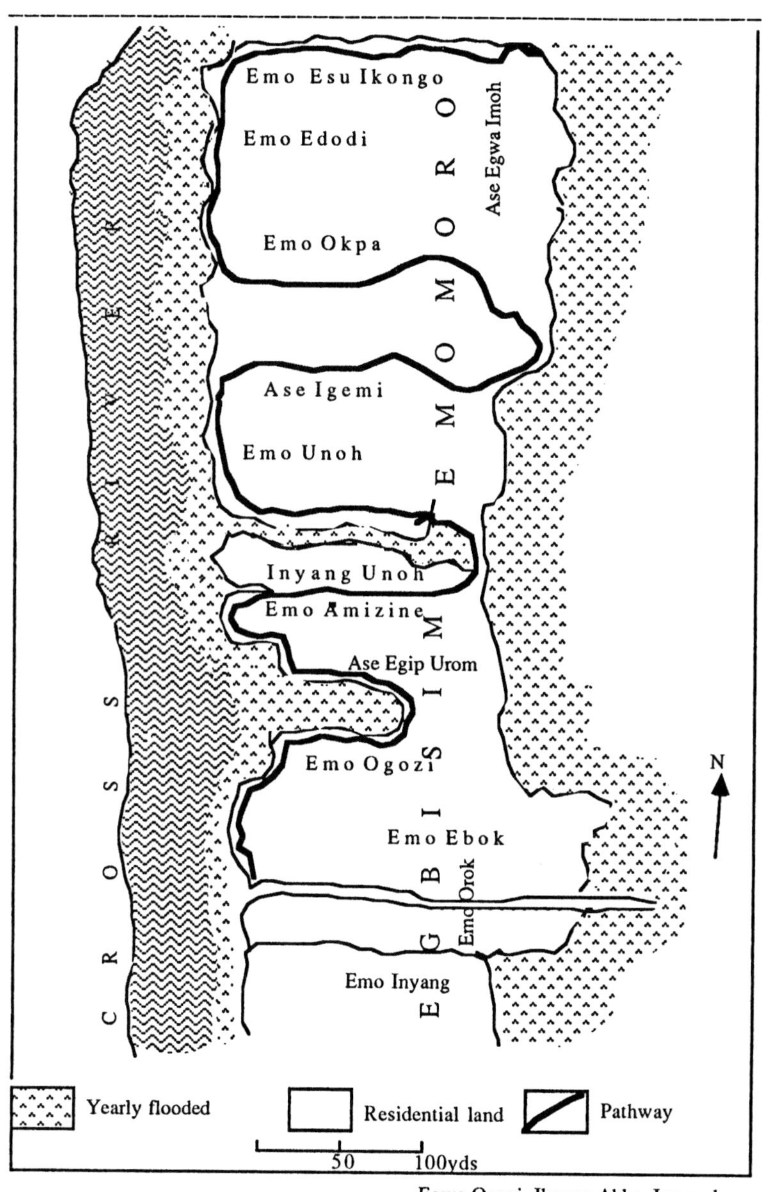

FIGURE 4. Odumugom-Agwagune, cluster of villages

is a result of landslides and denudation that have weathered the coast and continue to cut deep into the villages' space, clustering inhabitants into about forty three acres of space.

Abini, to the east of Agwagune, is located in relatively safer surroundings on a slope that offers good surface for runoff water. Its geology, a mixture of sand and stone, provides firmer earth, so that individuals rarely experience the problems with mud and flooding that hamper movement between and within the villages during the wet season. It is also located on the Calabar-Ikom highway, a major Nigerian highway offering easy access to principal Nigerian cities. The major ecological problem for Abini is lack of waterways. Two inland streams, Ubam and Udip, meander through the surrounding forest lands of Abini and Agwagune collecting into large bodies of water in the rainy season but drying up completely in places during the dry season. These inland streams are the main sources of water and also serve as fishing grounds; however, fishing is economically less important in Abini than in Agwagune.

Many Biase villages experience severe leaching and have a monthly mean rainfall of about twenty-eight inches. These have resulted in very poor soil that can support only such staples as cassava, cocoyams, yams, plantains, maize, vegetables, and a variety of other marginal cultigens (table 2).

Since there are few exploitable resources to attract government interest in the community, there is a general lack of government-provided amenities such as modern health facilities, salaried jobs, passable roads, a good sewage system, modern transportation, electricity, or a clean water supply. Abini and Agwagune typify the isolation, poverty, and political helplessness characteristic of all Biase. The lack of an all-season road, sufficient vehicles to run usable routes during the dry season, seasonal flooding, and the lack of a market are some of the reasons given by the Biase as responsible for their undeveloped economy. The major links

TABLE 2 Importance of Major Food Crops

	Cassava		Obiara		Yams		Maize		Rice		Plantains	
Village	Yes (%)	No (%)	Yes (%)	No (%)	Yes (%)	No (%)	Yes (%)	No (%)	Yes (%)	No (%)	Yes (%)	No (%)
Emomoro	100	0	100	0	60	40	73	27	0	100	20	80
Egbisim	100	0	100	0	54	46	86	14	6	94	6	94
Abini	100	0	100	0	56	44	74	26	47	53	20	80

Note: N = 30.

between Odumugom villages and the outside world are a nine-kilometer dry-season road that connects the villages with the Calabar-Ikom highway and the Cross River, which links it to neighboring coastal villages.

Usually all the land surrounding Agwagune is flooded during the rainy season from June to the end of October. At this time, waterlogged, mud-filled routes provide the only access to farms (plate 4). The nine-kilometer road is barely usable during the rains. About six kilometers of it, coming from the Adim end, affords minimal passage for vehicles, but there are large broken and mud-filled sections stretching for over fifty yards in some places. Two wooden bridges break down regularly and put the road out of use for months. People report that the condition of the bridges threatens the lives of travelers and the safety of vehicles that try to get to Agwagune villages. The last three kilometers is totally flooded and passage here is possible only by the use of small and medium-sized dugout canoes. Many people either swim or wade through flooded sections of the road because there are rarely canoes to ferry people across.

During the rainy season the Cross River is swollen with water, overflowing its banks at certain points and flooding some coastal communities. The river route gives easy access to more villages at this time, and travelers use small and medium-sized man-powered and motorized canoes because it is fatiguing to walk the long distances. The farm routes are also totally covered with water, which makes them less difficult to use, because it is easier to go longer distances by canoe than to wade through mud and water. We experience the most difficulty going to our farms just when water begins to flood, and again as floodwater starts to recede, because at these times the routes are mostly covered with mud.

A familiar experience is for entire villages to be flooded. For three weeks in July 1990 Agwagune faced the danger of being submerged when the Cross River overflowed its banks much earlier than usual. River water flowed into the villages, necessitating the use of canoes as the main means of moving about. People feared the loss of their homes and the destruction of their crops, planted only a couple of months earlier. But the water later receded and the land became dry once again, so people could move about with less hindrance. Before the end of October, the river itself started drying up and sandbars began to appear. By the middle of November, the flooded bush dried up, the rain ceased, sand spread far and wide over the entire coastal area of Agwagune villages, and people once again began the familiar routine of walking about nine hundred yards across a wide stretch of sandy riverbed to fetch drinking water at the far shore. Fishermen who had been waiting in the bushes for another flood to bring in more fish folded up their gear and returned to the villages. The dry season had begun.

By the end of October, the dry season has set in completely and the intense heat of the sun dries up much of the flooded land, including travel routes. But the road offers only minimal access because of the presence of large ravines and gullies from past rains. Men from the neighboring suburban town of Ugep occasionally maneuver their cars through this road to pick up a few passengers for a fee. So for a few months in the year, between October and March, the Biase see one or two cars in some of their villages. But since there is no regularly scheduled transportation service, most people cover the travel routes by foot.

The Biase regard their lands as the exclusive property of villages which depend on the support of community members for the maintenance of territorial rights. Territorial boundaries are clearly marked in some places with huge concrete pillars constructed by the Biase in the early 1900s, or with trees; no village land may be given away without the collective sanction of all patrilineages in the village. No patrilineage, for instance, will lease or loan its land to another community without the consensus of its members. As I describe in the next chapter, however, the preservation of territorial rights has come at a high cost.

Technoenvironment

There has been little change in technology over the past two hundred years. But the minute changes that have occurred have had a significant impact on strategies for meeting the environmental challenge.

The Biase are primarily horticultural, with all the attendant characteristics noted long ago by Pelzer (1945): rotation of fields rather than crops, use of the hoe, clearing by fire, reliance primarily on human labor, and absence of draft animals. Stakes, spears, hoes, and machetes are still the primary tools. Although flintlock guns were added to the repertoire before 1900 through trade with the Efik to the south (Ubi 1986), not everyone owns a gun, and guns are associated more with masculine pride than with hunting. Less than 2 percent of Emomoro/Egbisim men gave hunting as their occupation. The knowledge of how to construct temporary bridges (plate 5), canoes, houses, traps and snares, of how to craft ceremonial materials, and of how to discover ranges and trails of animals have been perfected and amassed into a repertoire of survival strategies in a difficult environment. For over two hundred years, the Biase were relatively self-sufficient, producing an agricultural surplus which they sold to the Efik, the Afikpo, the Arochukwu, and other communities up the Cross River until circumstances resulting from the Nigerian civil war thirty years ago engulfed them in a national food crisis. Later we will see how their consumption and production patterns changed in response to the city experiences of returning emigrants and young men who had joined the army or had otherwise been isolated from the village.

Cassava is the major food crop we process regularly. People refine cassava into *gari*, a staple food in many parts of Nigeria, particularly in the south. The traditional manual method of grating cassava is being gradually replaced by petroleum-powered cassava-grating machines which are available for a fee in many of our villages. Women dig up the tubers, peel off and discard the skin (plate 17), grate the cassava by hand or by machine, sieve off the fiber, dry it, and toast it with a little oil in wide iron pots until it is loose and grainy in texture. Gari, the final product, is stored in either fifty-kilogram bags or large enamel bowls that will preserve it for up to a few weeks for household use.

We gather some wild fruits to supplement the vegetables grown in home gardens. A wide variety of fruits and plants which are important food sources grow naturally in the surrounding bushes. We do not usually restrict access to naturally occurring foods unless they are on people's farms, in which case their owners may sell or give them away on request. There is about a 30 percent reliance on fishing for food; nets, hooks, and spears are used. Along with the round-rimmed nets used for *bob* (pond fishing), all fishing implements are obtained through trade. Hook lining is done along the riverbanks and ponds, mostly by older children and adolescents who do not yet own canoes. Some use *asanga* (a stringed hook suspended from a piece of bamboo, baited and dropped in ponds). Very few fish are caught by this method, but it gives children good practice.

Adult males bring in large catches using *meri-meri*, a technique borrowed from Afikpo fishermen involving the use of about a thousand hooks strung on a piece of string. Without any bait, meri-meri hooks are stretched across strategic points of the river or stream. As fish pass by, they become entangled in the myriad of hooks and are caught. Men stretch nets across certain sections of the river or pond to trap fish; a few cast nets. The Biase, however, do very little river fishing. The spear is not a major occupational tool, although spearing is done during communal pond fishing. Fish spearing is often discouraged in small ponds because people have been accidentally injured.

We do very little hunting, although most adult men own hunting guns. Hunting is often done at night by men who use carbide-powered head lanterns to find their way and to search for game. We also trap animals on a small scale using locally made snares. Not much hunting is done by either method, as there is very little reliance on meat; fish is almost always available for food. There are only a few artisans in Biase. Such craftwork as weaving floor mats, fishing nets, and carrying baskets is infrequently practiced.

As in most southern Nigerian rural communities, the Biase live in wattle-and-daub rectangular houses built with the combined effort of men and women. Community members often organize work parties to help build houses for friends and relatives without charge. The owners of the project often provide food and drinks. Previously work parties were more prevalent, because young men and adolescents were available to perform such tasks. Now, adolescents have gone to schools in the cities or have migrated to seek wage employment. In recent times the spiraling need for cash has affected cooperative work; people are increasingly selling their labor. Construction costs vary depending on the type of house being constructed and the reputation of the workers involved. Some people have constructed permanent buildings of cement and corrugated iron sheets in place of our traditional wattle-and-daub houses.

Apart from the wooden part of their work tools, the Biase import every item they use in their various occupations—guns, machetes, spear points, and hoe blades. Some men and women know how to design the wooden handles for hoes and machetes, which are the two most important tools (plate 13).

Transportation depends largely on human energy. During the dry season we carry items on our heads or in locally woven baskets balanced on the head, and walk an average of six kilometers to farms and markets. Small dugout canoes are important for transportation in streams that offer minimal passage in dry and rainy seasons. As children we learn early to manipulate the dugout canoes some of us row to our two-room village schools or to go fishing during the rainy season. We commonly wade through water and mud-covered trails to our farms.

In the absence of electricity, kerosene lamps are still the only means of illuminating homes at night. At the time of my study there were a few radios and two working television sets in Odumugom and ten in Abini, all powered by generators. The two functioning electric power generators in Odumugom supply single-point light to a handful of the households of friends and relatives of the owners. In some villages people still often bump into each other at night as they take evening walks along the dark trails that connect settlements within villages.

Economy

The allocation of productive goods—farmlands, fishponds, and fruits—is controlled by *etoima* (heads of patrilineages) who allocate farmlands to the *ima* (lineage members); the ima jointly exploit the fishponds or gather the fruits. Movable items such as tools for exploiting resources usually are inherited within the matrilineage. But inheritance of most property is minimal because items rarely survive long enough to be passed on to others. Fishing nets get torn within a season or two and are discarded, spears with broken handles are put away for years without being repaired, machetes lose their handles, guns malfunction and are too expensive to repair, wattle-and-daub houses fall into disrepair. Except for some farmland which belongs in the matrilineage and ceremonial materials such as pots, clothes, body ornaments, and musical instruments that are collectively owned by matrilineage members, very little material of economic value is passed on to the next generation.

Land distribution starts in either January or February as the patrilineages prepare for the next farming season. The pattern of farmland distribution in Biase varies between villages and within lineages. In some villages allocation is determined on the basis of neither age nor sex, while such criteria are important in others. In my subgroup, the Agwagune, the Onun of our patrilineage simply announces a day for everyone to gather in the lineage house from where we all go to the area designated for that year's farming. On arrival, men and women in turn use visible markers to indicate the amount of land they want to farm. Some of us who want farmland but are unavailable at allocation time are assigned lands in our absence. In some lineages elder males mark out and assign plots of farmland to individuals on the basis of need and their ability to work the plot. If anyone is left without land at the end of this exercise, the Onun readjusts the apportionment to accommodate whoever was left out. The primary aim is to make sure that everyone has an adequate portion of land to farm; no one has ever been reported to be left without farmland.

Market exchange, *afia,* is important in resource distribution. The Biase traded with the Efik of Calabar long before Europeans arrived in southern Nigeria (Alagoa 1971); canoes have been an important economic and political resource for such encounters and for the political control of the upper reaches of the Cross River. Yams and palm oil, confined within the sphere of male production, were the main trade commodities. Other goods and services were shared by reciprocity, redistribution, and money exchange. Long after the Europeans arrived, Biase forest land remained lush and rich in resources because it was minimally exploited. With our simple implements of stakes, spears, hoes, and machetes we produced surplus food. Trade was an important source of exchange involving heavy traffic in goods and services along the Cross River route. Elders, assisted by traditional secular and mystical associations, were feared and respected, and conformity to social norms created a stable social environment. The Biase consolidated a social structure that brought them attention not only from other southern Nigerian groups but also from colonialists who, confronted by potential rivals, challenged the Agwagune for control of Cross River coastal communities.

Although the more common trade form is periodic marketing (*ikpo*—named after the designated weekday) which operates on a four-day regime, an active informal market involving exchange of food and nonfood items goes on daily in every community. Reciprocal exchange also exists to some degree, and occurs regularly among members of the same age set, within groups of friends, and sometimes between families.

We organize work on the basis of *ba-aka* (the kindred), *eromo* (friends) *anerom/aneba* (mens/womens/groups), *egop* (age sets), and *ikpafini* (thrift associations), but these units vary in importance. Although the family is the preferred basis for organizing labor, other nondomestic units have become more important because family members are increasingly involved in alternative economic interest, such as wage labor outside the community. The more important means of organizing work are eromo, which perform work for each other reciprocally and sometimes for other family members. However, most eromo work groups sell their services to any person (including members of their families) willing to pay the required fee.

Despite the inconvenience of the rainy season, we all await its coming with much excitement because it marks the beginning of the fishing season, which brings in more income. All kinds of fishing activities begin in *eroi*, the bush, which is flooded by the rains and the overflow of the Cross River, as men fill streams and rivulets with hooks, nets, and lines. There is often a decline in the availability of fish following the end of *bob* (fishpond) season in April; but in June the selling and consumption of fish take center stage once again.

Women's labor in the rainy season is much in demand because

women do all the weeding, which is the main farmwork at this period. Work usually accumulates in the rainy season due to the difficulty of getting to farms, because many women cannot find canoes to ferry them to their farms. Sometimes it rains so hard that women cannot go to their farms for three days or more. This is frustrating because the weeds accumulate in their absence.

Men's work seems less strenuous. We leave in our canoes to go fishing every morning, return about two hours later, spend much of the afternoon just sitting and chatting in the lineage centers, and return to check our catch in the evening. Often we build small huts in the place where we fish and stay there for up to a month, drying, selling, and eating an enormous quantity of fresh fish. Occasionally we row our canoes out to our farms to direct climbing yams shoots up their stakes.

Apart from fishing, the Biase have very few off-the-farm occupations. Reports of the earlier presence of iron forging, mostly from informants' accounts, have remained unsubstantiated. For the past fifty years, such work tools as hoe blades, spear points, and machetes have been obtained through trade with the neighboring Igbo people of Abiriba, Arochukwu, and Afikpo. At the time of my research, there was an Igbo blacksmith working in Odumugom mending iron pots, fixing locks, and sharpening machetes.

Because Biase roads are mostly unusable for about seven months of the year, conveying food to neighboring markets is a major challenge. Women carry food on their heads from farms about six kilometers from home, storing them in their homes or in transit huts until they can be sold locally or taken to neighboring markets.

Raising domestic animals is popular. Seventy percent of the households in my census own at least one goat or chicken. She-goats are preferred because of their reproductive advantage. Goats and sheep are sold at very high prices for festive or ritual events; rarely do people kill and eat their domestic animals for nonceremonial occasions. But we always have something to celebrate—a marriage, a birth, a naming, a death, a visit, a new bicycle, a new radio, building a house—all call for some form of celebration.

Social Organization

The Biase have both matrilineal and patrilineal corporately organized descent groups, often labeled "double unilineal descent"; in southeastern Nigeria we share this uncommon system with the Yako (Forde 1964) Mbembe (Harris 1965), and the Afikpo (Ottenberg 1968). We live in small patrilineage groups which cluster together into large local patrilineages or patriclans. These large patrilineages own land corporately, organize farming rituals such as those that mark the new yam festival, enforce rules, arbitrate disputes, and prescribe punishment for social infractions. We all trace relationship to a single ancestor, and in spite of the genealogical link to an apical male ancestor, refer to ourselves as *ba-aka,* "children of the mother." This term also designates sisters and brothers.

My grandfather was the previous head of the household now headed by my father. My father's brothers, sisters, and cousins lived in the same household or in adjacent households. Being the eldest child, my father inherited the headship of the household, while his younger brothers moved into adjacent households or built their own homes in the patrilineage compound. My father addresses cousins or brothers as "brother," and the rest of us address them as "father." When work gets light in my father's farm, any of my other fathers may request my assistance at farmwork.

Male membership in our patrilineage is relatively stable partly because we marry exogamously; our wives and children reside with us in the patrilineage compound. Men retain their rooms or living quarters even after they move out or set up homes elsewhere. During my prolonged absence from Agwagune, my living quarters are occupied by my younger brothers, unmarried sisters, and divorced or widowed aunts. Some of these people may also decide to live in the adjacent households of my father's brothers or cousins. My patrilineal cousins do not have less rights in the patrilineage than I do, although they may informally defer to me in some cases because of my education. My educational status, however, is never an important factor in my social relationship with my brothers. This is a little to my advantage because my fellow villagers usually equate my education and long absence from the village with the loss of local skills. I am their favorite target when they talk about how naive in traditional ways education has made some of us.

There are fruit trees that belong to the parilineage, but my father also owns fruit trees that he inherited directly from his father. He would often take me to the bush to show me some of the trees that would be handed down to me. Insofar as I am physically present to harvest my fruit trees, no one harvests the fruits without my permission. In my absence, however, any male in my patrilineage may request temporary ownership of the trees unless my father is able to tend and harvest them.

I also have significant ties to my mother's matrilineage. Like my mother, most of the females in my matrilineage are scattered over various patrilineages or live outside the village. But we often assemble in the matrilineage *aka edong* (big house) for a lineage meeting or to celebrate some event. I see little difference between my relationship with members of my patrilineage and that with members of the matrilineage. Everyone recognizes my rights through my mother to other members of the matrilineage. The siblings of my mother and grandmother are my fathers or mothers; I address the children of my mother's siblings as "sister" or "brother." But I inherit very little from my matrilineage. When my grandmother died in 1980, I was angry that all her jewelry went to my sisters; I felt some should have been given to me for my future wife. In my anger, I asserted a claim to my grandmother's case of china and moved the case to the safety of my father's house. My aunt still constantly reminds me that the case should be returned to where it belongs.

Although I was deprived of my rights to my grandmother's accumulated wealth, I was confronted with a situation where I had to settle my grandmother's debt. During my research, about ten years after my grandmother's death, a man in my matrilineage approached me to demand the repayment of some money he said my grandmother had borrowed from him to pay my school fees while I was in high school. Nobody in my matrilineage remembered the debt and my mother advised me that I should investigate it thoroughly before deciding either way. However, I repaid the debt promptly because it seemed awkward that I should continue to farther my education at this man's expense. It would also be embarrassing if people learned I was haggling with the man over a debt that my grandmother may indeed have incurred. The point is that the corresponding rights of economic obligation that belong to women of the matrilineage are renegotiated on the basis of who is best able to repay a debt independent of the matrilineal category.

I have unrestricted access to the farmland in my matrilineage and have sometimes been warned that the amount of land I was demanding would not be enough to feed my mother and my siblings after I left the village. I am often encouraged to secure as much land as I can comfortably farm. Just like the members of my patrilineage, members of my matrilineage expect me to devote adequate time to the work demands of the lineage. During my research, my patrilineage set a date for everyone to work on the farm of a lineage Onun on the same day that my matrilineage was going to construct a wooden bridge to the farm route. I was told that since I could not be at both places at the same time, I should hire someone to represent me at one of them. I hired someone to work for my patrilineage and went to join the bridge construction crew. An alternative solution would have been to send food to the workers at one site and be present at the other. The penalty for neglecting either lineage is the assessment of a fine or exclusion from lineage privileges.

Women of my matrilineage also build their own houses either in the compounds of their matrilineage or in other compounds. When my mother's sister built her house, I was surprised that it was not adjacent to my grandmother's house or even in my matrilineage compound. My grandmother explained that my aunt was entitled to build a house in any of the matrilineage compounds if the leaders of the compound granted her permission to do so; she pointed out similar residences in the matriclan compound. Like my aunt, other divorced, widowed, or older women either live alone in their family homes or share residence with other matrilineage members. Individual members of my matrilineage have rights to my services and do not hesitate to request that I go with them to farm or help out with chores around the house.

Irrespective of a child's kinship affiliations, any adult in the village who finds a child misbehaving is expected to discipline the child, usually by smacking the child on the buttocks with his or her bare hands. My grandmother severely berated a woman of another matrilineage for failing to discipline me when she caught me climbing a coconut tree belonging to her lineage. She gave me a good whipping and complained to everyone in my matrilineage that the woman was abetting my unseemly behavior.

The descent system briefly discussed here has been reconfigured in certain respects, although the basic principles of residence, relationships, inheritance, and familial expectations still exist. Postmarital residence is increasingly neolocal, a few people are changing their lineage affiliations by formally severing ties with their birth lineages, inheritance of movable properties by females of the matrilineage is being increasingly challenged by adventurous males, and some lineage members have become less obliging in offering unrestricted assistance. On the whole, however, the double descent principle still operates.

The Biase do not have a specific age at which a girl is considered ready for marriage or must get married, but marriages rarely occur before a girl's sixteenth birthday. Before a girl is married, her parents prepare her for *ugim*, which she undergoes in a room in any of the homes of her patrilineage. Although this practice is commonly referred to as fattening the bride, *ugbugba ugim* (the ugim maiden) does not have to be a bride, nor is it a requirement that she must get fat. It is the custom that ugim must precede pregnancy, so every girl is expected to undergo ugim before marriage, but pregnancies outside marriage sometimes occur. When they first learn that their young daughter is pregnant outside marriage, some parents may express their displeasure by whipping the girl for her apparent waywardness. After their anger is vented, the family begins preparations for their pregnant daughter's ugim, and the girl's mother continuously monitors the health of her daughter as family members excitedly await the birth of the new child; chastity, however, is not greatly treasured in Biase.

In anticipation of an accidental pregnancy, an unmarried girl undergoes ugim as soon as she is suspected of being sexually active. Her ugim takes place in one of the rooms in her father's house. The ugim of a bride takes place in the groom's home. While she is in ugim, a girl's parents, friends, and relatives continuously parade large bowls of pounded yams in and out of her room, because the objective really is to make her gain some weight. But many girls eat very little of the food they are served. When we were growing up, our favorite spot was the ugim home, where food was never in short supply. The ugim girl would sometimes ask me to clean the bowls and return them to their owners. This was never before I had thoroughly stretched my stomach with meat, fish, foofoo, and yam meals that the ugim girl was unable to eat. My grandmother said that girls used to remain in ugim for as long as two years; Now few stay longer than a month. Some of our girls return from college and fit their ugim into a one- or two-week school vacation.

Marrying a wife, *onegwa*, is an elaborate drawn-out event involving the transfer of palm wine, food, and a small amount of money from the groom's family to the bride's. Negotiations between the families of the groom and the bride are marked by humorous accusations that one is seeking to take advantage of the other; in many cases, the two individuals in the relationship have already come to an agreement and negotiations are rarely forestalled. Most arranged marriages prosper. Although it is no longer a dominant practice for our parents to arrange our marriages, family members still play a central role in marriage negotiations. The family of the groom offers the bride's family food and drinks at various stages of the negotiations. When matters are settled, the family of the groom consults with his group of friends and sets a date for capturing the bride. On the day set for the capture, the women of the bride's family decorate her body with white and red chalk, tie a standard locally designed cloth around her waist, put an ostrich feather in her hair, and fit bangles made from elephant tusks around her ankles and wrists. Shortly after a few more ceremonies, the groom's friends arrive, fire their flintlock guns into the air, and burst into the room where the bride is held. As all the women in the house and everyone gathered outside shout in excitement, six or more of the men quickly grab the bride, lift her onto their shoulders (two men support her shoulders, two, her waist, and two, her legs), and run with her all the way to the groom's home. The bride's family meanwhile engages them in joking abuse but does not try to stop them.

Men, much more than women, express a preference for polygyny (*anebaogwu*). Men with pregnant wives say they need a second wife to help with farm and domestic work; those with older wives need a second woman to assist their elderly wives with their labor. The majority of men have only one wife, although there is usually a concubine (*ovuk*) on the side. Men nourish the relationship with their concubines with occasional gifts of fish, secretly supporting their concubines farmwork by hiring labor for them, and sometimes giving them money. Although extramarital relations are common knowledge in the community, no wife who is aware of her husband's extramarital relations accepts them. Unmarried men and women freely take lovers without communal sanctions.

Depending on where Agwagune children are raised, they may grow up and bond more with either the patrilineage or the matrilineage. Since I grew up in the compound at my matrilineage, I ended up establishing primary membership in my matrilineage. This never excludes me from any of the rights and privileges of my patrilineage, nor is my mother's brother more significant than my father or my father's brother; I relate to all equally. In the same year I may ask for farms from both lineages without encountering any problem. However, I have fewer inheritance rights in my matrilineage than the sons of my mother's brother. For instance, I would not be considered more favorably than my cross-cousins for succession to the Onun stool if my matrilineage were the ruling lineage. But in the absence of a male acceptable to everyone, nothing bars me from being nominated by my matrilineage. At all times I am strongest in my patrilineal relations, where, although I have not established formal membership, I am recognized as a son of the patriline. In matters of inheritance or succession to the stool of Onun Agwagune (in the case of Inun patrilineages), I will be considered strongly in terms of my birth position in the patrilineage.

But the pattern of inheritance differs in other Biase communities, as in the case of the Abini, where matrilineality is important. On the death of his father, a man is not entitled to such property as farmland and fruit trees outside his father's physical dwelling. Such property reverts to the matrilineage but is passed on to other families in the agnatic line. A man may then inherit property from his patrilineage, but not directly from his father. Thus, although the Abini inherit matrilineally, patrilineal inheritance is not absent. People inherit farmland from their matrilineal kin, but property within the physical dwelling may be inherited patrilineally, or from father to son. Among the Agwagune and the Abini, the eldest male is usually expected to have a controlling interest in property passed down the patriline. However, this is often not the case; a resolute younger man or woman may assume control.

Male children are preferred because men perpetuate the family name, but females occupy important social and economic positions in Biase. Some women exert a profound influence on community decisions, sometimes because of their education and kinship relations, but more often because they have been assigned authority by other women. The Agwagune women's association, for example, assigns certain duties to its members which makes them highly regarded by other community members.

Stratification is indicated by the presence of ruling houses (*ugom inun*) that traditionally provide candidates for village headship. Other families usually provide the leaders of such social groups as *egop, abu, ebrambi,* and *ekpe.* Stratification results from differences in income and property ownership. Some farmers are more prosperous than others, and a few forest resources are also privately owned. Most people, however, are within the same socioeconomic group.

Age grades are structurally important in consolidating social relations, performing communal tasks, generating income, enforcing social rules, and affirming ideology. Intercommunal cultural exchanges, such as invitations to feasting or to launching a new project, keep age sets within and between communities in touch with each other. In Agwagune, age sets ironically go by such foreign names as Unity (ages 25–30), Freedom (31–35), Tunisia (36–40), Guinea (41–45), Congo (46–50), Ghana (51–55), Ibadan (56–60), Benin (61–65), Africa (66–70), French (71–75), Bonima (76–80), Company (81–85), America (86–90), Sierra Leone (91–95), Obioko (96–100), and Etan Agaghara (100+) (table 3). Significantly, only the two senior sets go by indigenous names. Onun Aquah of the Sierra Leona set said that all age sets went by local names when he was growing up. Members of the other sets said they chose their name without thinking about its significance. It is possible, how-

TABLE 3 Agwagune Age Set Chronology

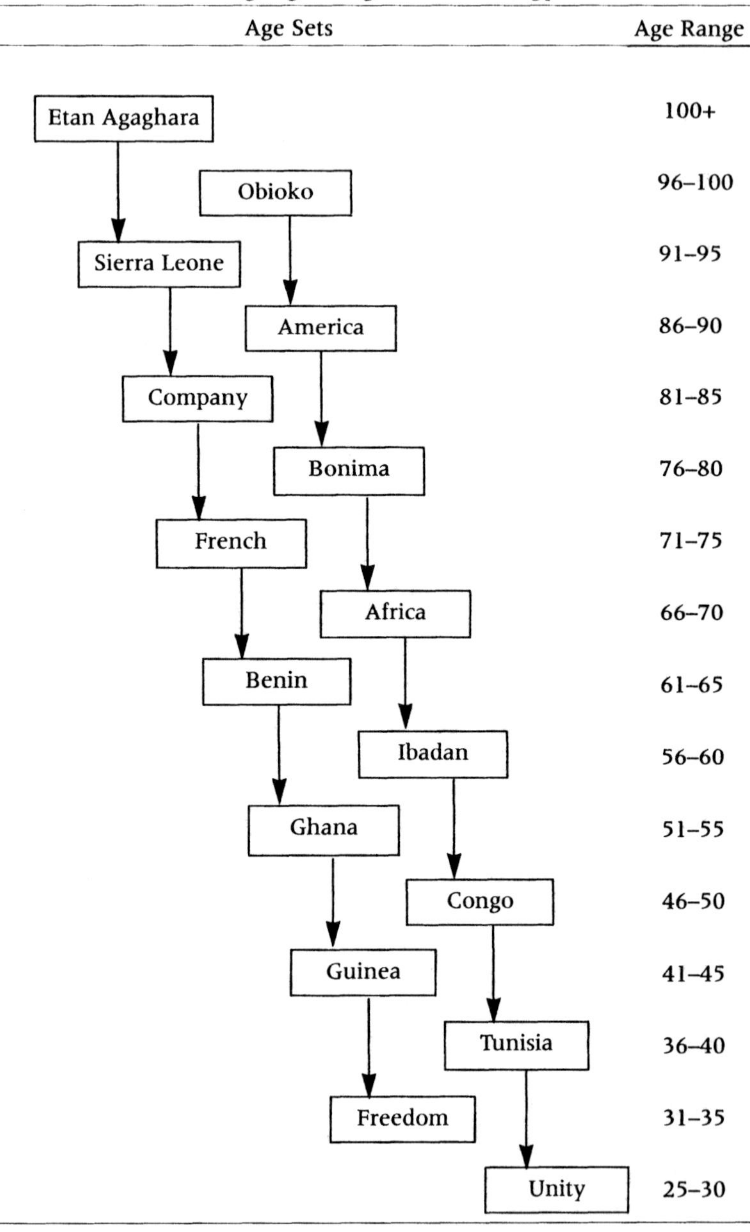

Age Sets	Age Range
Etan Agaghara	100+
Obioko	96–100
Sierra Leone	91–95
America	86–90
Company	81–85
Bonima	76–80
French	71–75
Africa	66–70
Benin	61–65
Ibadan	56–60
Ghana	51–55
Congo	46–50
Guinea	41–45
Tunisia	36–40
Freedom	31–35
Unity	25–30

ever, that the precedent was set by some Agwagune men who were introduced to those names during their first contact with Europeans and during their service in the World Wars that took them to countries and cities outside Biase.

In the Biase age grade system, members are permanently locked into their sets; as older sets disappear or move up in the social hierarchy, younger sets replace them and new ones are named. New age sets are formally constituted when members are about the age of twenty, although children of a prospective set start meeting informally when they are much younger. Members of each age set are classified as children of the higher alternate age set (see table 3); the age set "parents" traditionally assign a name to their children's set in an elaborate ceremony featuring the presentation of food and drinks by the prospective age set to their parents in the hope that a "suitable" name will be assigned. Each age set is thus a semiautonomous body which follows traditional rules in establishing itself but has autonomous rules for regulating its activities.

The age sets from French upward are composed of older men and women who are exempted from participation in organized communal labor, although members may participate if they choose too. Their role as advisers and as the repository of communal knowledge, which is often the basis for making rules and decisions, is important in Biase.

An important aspect of social groups in Biase is that they maintain continuity in membership. Sacred societies such as Abu and Egup, age sets, friendship groups, and secular associations operate on the basis of permanency; people do not resign from their social groups even if they no longer reside in the area. Among friendship groups, however, certain members may be dismissed, while others may withdraw for any reason from the disorderly conduct of some members to the rebellious attitude of others. But generally continuity is maintained in the membership of friendship groups. The functions of your friendship group include capturing your wife or rating a prospective husband, helping to construct your house or supplementing farm labor, organizing social events for members, and playing culturally designated roles at the birth or death of family members. Friendship groups comprise primarily members of the same age set and sex, although older members mix with younger members in quite a few others. Many people belong to more than one friendship group but must take the time to honor their obligations equally to all their friends. People who are absent when group members have obligations are represented by other persons in their families. For instance, whenever I was away from Agwagune, either of my parents, my grandmother, my brothers, my sisters, or anyone in my patrilineage or my matrilineage would represent me in any matter concerning my friendship groups. There is no restriction as to the sex or age of the representative.

Mystical and secular associations are the major institutions that regulate social behavior, assign rewards and punishments, and influence kinship relationships. Members of mystical associations are feared and revered by others in the village because of the supernatural powers associated with the mystical associations and because of the fear of supernatural retribution for anyone who neglects association dictates. Although such male associations as Abụ and Ebrambi and the female association Egup have socializing and judicatory functions, mystical associations such as Inyono and Ekpe are said to be useful for individual members rather than members of their kin groups. Members of the Inyono association, for example, are said to maintain their standing in the association by cabalistic "sacrificing" of close kin to fulfill the demands of the association. The Inyono association of the Biase exists in various forms among other southern Nigerian groups and bears some similarities to the Ogboni of the Yoruba in terms of the special political and economic privileges members enjoy (Sklar 1991: 16). Membership in any of the mystical associations is aid to be transferable from one community or ethnic group to another. On the other hand, membership in the secular associations is restricted to persons of the villages. Some secular groups, however, admit settlers who have shown a commitment to communal life through farming, marriage, participation in work groups, or other forms of lineage affiliation.

Some of the most important associations among the Agwagune and Abini are their Aka-e-Mitin Aneba (women's great meeting; see plate 19). While they have little political authority within the towns, they run a powerful government within their caucuses, hearing and settling disputes among members, exacting fines, and in the process accumulating a strong financial base. The community perceives the women's organizations as the ultimate source of emergency financing, and men regularly go to them to request loans to settle communal debts. The importance of the women's associations is better understood in the context of the importance and eventual decline of Abu, the premier society of the men.

Abu is the preeminent male association of the Biase, although the intensity of the observance of its rules and the practice of its rituals vary among Biase groups. In precolonial times and until the late colonial period, Abu served as the governing institution of the Agwagune, enforced social rules, prescribed social decorum, and proscribed certain behaviors. Under the command of Abu elders, the Agwagune enjoyed unquestioning loyalty and obedience from young men and women, who were careful not to contravene the numerous proscriptions of the Abu association and suffer the severe judgments of the association. The public symbol of Abu is the frond of the palm tree. Abu members may tie them to objects or simply place them in a particular area from which they want everyone to keep away. This Abu symbol is understood by everyone in the community to mean that whoever violates the indicated proscription is liable to Abu sanctions. Although it is strictly a male association, its primary responsibility is to uphold communal ethics rather than to protect men. Membership in the association does not save an offender from punishment.

When the Agwagune man exclaims "Ebiabu!" it is an affirmation of his resolve to accomplish a particular objective, an endorsement of a mutual agreement, a determination to enforce a rule or punish a contravention. Ebiabu! is the exclamation that identifies a member of the Abu association and must never be uttered by females or uninitiated males. Every Agwagune male becomes a member of the association by initiation during the Abu festival, which occurs every seven years.

The Abu initiation is an affirmation of maleness rather than a transition to manhood; initiates are usually children between the ages of four and twelve comprising about two age sets. For seven days, all of us male children are secluded deep inside the forest that serves as the Abu domain where initiation rituals are performed. During the initiation period Abu elders teach us appropriate social decorum for associating with community elders and a variety of skills needed in a culture where we start when still young to contribute to the support of our families. At the end of the initiation, we are instilled with fear for Abu secrets we are too young to understand, we identify with the preeminent position of the association among the group, learn to fear and respect our elders and value their wisdom, practice behaviors that help us to bond with our peers and group members, and master the caution we need to deal with strangers. To announce the initiates' new sense of responsibility to the community, the new initiates end the initiation with a procession, singing and dancing around the village and family compounds. For the first time in many days, our mothers see us again as they proudly watch us displaying the intricate Abu dance steps we learned during our time in seclusion.

The Abu hierarchy is closely tied to the age grade system of the Agwagune. As new Abu initiates we are unimportant in the political structure of Abu, although we can participate in most of its activities. As adolescents, we constitute the Agop Oge (those wielding knives) and have the primary task of performing menial tasks around our villages. We attain political recognition between ages twenty-five and thirty-five, when the two age sets that fall within that range are recognized, in the year of the Abu festival, as Agop Ibe (those willing to die). We hold the Agop Ibe title for seven years, during which we enforce the rules of Abu elders. We are the first to confront any threat directed at the community and the first to be called out to perform any communal labor. We are elevated to the Ibuot Eto grade at the next festival, and the two younger age sets behind us become Agope Ibe. As Ibuot Eto we oversee the functions of Agop Ibe. We make laws, interpret social rules, arbitrate in all matters brought before the Abu, and pronounce judgments. Seven years later, we become Egot Uso, the last active grade in the Abu hierarchy. Having joined the ranks of community elders, our Abu functions now are only as advisers. As we get older we lapse into the Adum Abu set (old Abu members) and perform only ceremonial duties at Abu festivals. Because of the protection it offers community members, the fear and respect it commands, the quickness, fairness, and firmness with which it decides matters taken before it, and the severe fines it imposes on the guilty, the Abu association for many years functioned as the supreme government of the Agwagune. In chapter 7 I will discuss the events that led to the decline of the Abu association, the loss of the high position of community elders, and the emergence of the women's association, Aka-e-Mitin Aneba, as the most important social group in Agwagune.

Politics

We have three levels of Inun in Agwagune: Onun Agwagune of the entire group; the Onun of a village, and the Onun of a patrilineage. The two lower levels of Inun make up the council responsible for appointing the Onun Agwagune and constitute a ruling council in the absence of the Onun.

The highest level is the Onun Agwagune, who exerts much authority over the entire group and coordinates all actions involving the group and its neighbors. It is the Onun who makes the decision to go to war, to cease hostilities, to lease our land, or to initiate economic cooperation with other groups. No village in the group unilaterally enters into a serious relationship with another group without the approval of Onun Agwagune.

The appointment of an Onun in Biase is a complex process of searching for the candidate, "arresting" him, and installing him. It is a process that tests the patience of community members, the fortitude of the Onun candidate, and the wisdom of lineage Inun who have to formalize the appointment. When we say we are *searching* for the Onun candidate, we mean that we are waiting for the appropriate kingship patrilineage (for example, one of the two ruling patrilineages in Agwagune, Emo Imo or Emo Edodi) to name someone for consideration for the position by all the village Inun of the group. The names of possible candidates are circulated among the Inun and elders of the group, who secretly express their views on the choice. The search process may go on for a few months to several years and takes longer if the Inun do not agree on a candidate. Another reason why a search may take a long time is that there are some people who refuse to be appointed Onun and disappear from the group for many years, returning only after an Onun is installed. The Onun stool may be vacant for a long time if there is no acceptable person. In such cases some group elders may negotiate with a popular member of a kingship patrilineage who may be residing in a distant town to accept nomination to the stool. According to Onun Jonah Unoh Aquah:

> Once upon a time our people had gathered at a meeting with no Onun Agwagune present. I turned to the Inun of Agwagune villages and said, "Here is our Onun stool but no one is sitting on it. The government is going to seize the stool since there is no Onun." I turned to Onun Odim Edodi of Okurike and appealed to him to occupy the stool, but he refused, saying he would die if he became Onun because he was not from an Onun patrilineage. Your father, Uru Iyam, was the secretary at that meeting. Uru said we should go in search of Onun Agwagune. We came to Onun Odidi Okpitu of Itu-Agwagune and asked him to occupy the stool. Although he lived in Itu-Agwagune, he was originally from Okpatim Esu in Emomoro and was therefore entitled to be Onun. Okpitu agreed, and we went on to make him Onun.

If the Inun find and agree upon a candidate, they "arrest" him to prevent him from running away. The arrest involves the elders entering the person's house unannounced with the greeting, "Onun we salute you." As soon as he is so greeted, the elders surround him and place their hands upon him. The hand placing summons our spirits and ancestors to also arrest the candidate and make him unable to run away even if he wants to, but it mostly protects him from the evil intentions of others. The inun then set a date for the installation. The installation is marked by ritual sacrifices, songs, and dances by various Biase groups and even groups outside Biase.

Immediately below the Onun Agwagune are the Inun who head Agwagune villages. They are answerable to Onun Agwagune in matters relating to other villages in the group but may independently make decisions affecting people under their jurisdiction. The process of appointing village Inun is less complex than that for appointing Onun Agwagune. Although the patrilineage Inun still search, arrest, and install the candidate, there is less pageantry. Village Inun are nominated by the Inun of their specific lineages and announced to the rest of the group. The village Onun does not go on to become Onun Agwagune, but he will be considered as a candidate if he is of a kingship patrilineage.

Lineage inun make up the third level. Male elders of the patrilineage seek candidates from among the males closest to the founding member of the lineage and appoint one the lineage Onun. His primary responsibility is to supervise resource allocation in his lineage, to represent his lineage as well as offer support to the village Onun at the Agwagune council of Inun meeting, and to arbitrate cases within the lineage.

It is generally believed in Biase that the position of the Onun is linked to supernatural and supernormal powers. These powers are important in the Onun's discharge of his duties. They define his reputation among the Inun of other groups and command fear and respect among his constituents. Many Biase elders who were not Inun also shared this reputation. As a child in Biase, I was afraid to pass close to the Onun's house after learning of a particular spot beside the house of the Onun Agwagune. Anyone unprotected by supernatural agents who stepped on that spot would immediately disappear from sight and would reappear beside a dangerous stream miles from our village. Whenever the Inun and the elders were at meetings, my grandmother would warn me not to pass by the lineage center, because it would be dangerous to pass too close to the spiritual powers of all those elders gathered in one place. The association of the Onun position with the supernatural is one reason why some people do not want to be considered.

The Inun and the elders of Biase therefore commanded much reverence and respect and used these qualities to manage their groups without opposition from youths and with little dissent from the other elders. Earlier in our history the Biase had powerful Inun who respected each other's strength but deferred to the more powerful Onun of Agwagune, who had authority extending beyond his immediate domain (Nair 1972). Political organization revolved around the authority of elderly men who were respected and feared because their role as heads of patrilineages had supernatural sanction. Elderly men were able to manage the communities with little opposition; their orders were neither questioned nor flouted. The authority of village rulers extended to the control of land and all its manifestations, making it possible for everyone in the community to have access to good land on the basis of communal rules of apportionment.

This is not to say that rulers and elders were omnipotent. As has been noted for kings elsewhere in this part of Nigeria, much of their governance was accomplished through negotiation. But they were also able to influence events through their affiliation with feared associations such as Inyono and Abu (Harris 1965). Such a strategy is also reported for Kenya, where the custom of blood oath was used to control access to land and to determine ownership of land rights among the Mbeere (Brokensha and Glazier 1973, cited in Barrows and Roth 1990). According to Bromley and Cernea (1989), such traditional controls provide an effective organizational arrangement for sustained economic growth among rural peoples. Because people were not allocated farmland until they were married, more young adults still lived under the authority of parents, relatives, or family friends who strongly enforced social rules that governed appropriate behavior. The consequent tendency to be respectful of communal rules governing resource allocation and use encouraged behavior that preserved resources.

In Biase today, Inun exercise consensual leadership because authority is otherwise diffused among age sets and men and women's associations. The common complaint against the Inun is that they lack power and authority to represent their community in negotiations with the government, because they are weak and powerless and unable to manipulate power either within or outside the community. Later we will see how present-day dissenting voices and the breakdown of our traditional legal mechanisms have freed individuals to manipulate access to new political systems and the rights to land of other groups, resulting in a change in the character of territorial disputes. The older generation often refers to the days when the authority of the inun was firm as a standard for measuring the weakness of the present system and as an explanation for rural socioeconomic decline, which is often blamed on the new generation.

Ideology

The person you see first in the morning is important to the success or failure of your activities for the rest of the day. People rise before dawn to wake up their favorite *uso onne* (good face) people so they will see them first thing in the morning before they go fishing or to the market. Some people do not take a bath until they have completed the event for which seeing the good luck person is relevant. If you encounter a "bad face" person, you should cover your face and not acknowledge that person's presence. In such a case, you may acknowledge a greeting without seeing the other person's face. People who have slept in the houses of friends or concubines hurry back at dawn so they will be seen by prospective good luck seekers. Such caution is necessary for maintaining a balance between the real and mystical worlds, because nothing ever happens by chance. A man does not experience sickness or misfortune fortuitously; the wind does not blow nor the lightening strike by chance. People ask: Why did it happen to that particular man? Why did the tree branch break just then? The town priest, Onun Erot, always has an explanation. Shortly after I returned to the United States from the field, I managed to arrange to speak with my parents on the telephone in a friend's house. On that day I was recovering from a mild fever. My mother could not restrain her anger when she heard I was ill, because a few days earlier a man in my patrilineage had pointed his walking stick at her during a quarrel. We do not point our walking sticks at others unless we intend them some harm. My mother had complained at the time to our lineage Onun and warned that if anything happened to her or her children, that man would be held accountable. Somehow the evil intention of the man found its way to the United States and afflicted me far away from home. At the end of our conversation, my parents resolved to make a formal complaint to the patrilineage.

Attributing causality to other-worldly powers is a dominant theme in the daily life of the Biase, and the importance of these powers to individual and group survival is strongly respected. About 80 percent of the Biase profess membership in the Christian-based spiritualist church of the Brotherhood of the Cross and Star (BCS), founded by a Biase man but enjoying wide membership all over Nigeria and in other countries.

The remaining 20 percent are members of other Christian churches and adherents of traditional religion. Before the appearance of BCS about forty years ago, the Catholic church was dominant. People say there was also much witchcraft at that time, but its incidence has diminished since interest in spiritualist churches began to spread. Even people known to be strongly involved in feared traditional associations are now baptized in spiritualist churches.

The Church of Scotland Mission under Samuel Edgerley first arrived in Agwagune in 1876 (McFarlan 1957); in 1889, the Reverend James Luke opened the United Free Church in Emomoro (Talbot 1960). While many of us are nominally Christian, we find it convenient not to discard our traditional religion. Almost everyone practices rituals for acting out the strong belief in and worship of traditional deities. We have beliefs focused on the control of nature, reincarnation, veneration of ancestral spirits, totems, divination, sorcery, and witchcraft; we also practice magical rituals for protection or offense. Of the 269 households surveyed in a census of Emomoro/Egbisim villages, 235 household heads admitted practicing some form of traditional protective ritual; 229 of them also were members of Christian churches. Having been baptized in the Presbyterian church as a child, I have learned to carry a small copy of the New Testament in my pocket in the hope that the holiness associated with it and sanctioned by my Christian belief will protect me from all evil. In another pocket I carry a protective charm given to me by the priests of the Efa shrine, which protected me against enemy attack through all the months of my research. With both religions I order my understanding of the world around me and construct my responses to the challenges of life posed by the visible and the invisible forces that populate my worldview.

As the source of food and the home of ancestral spirits, the earth, *eziba*, is given more respect than other aspects of the environment. Our fertility deities have a significant effect on the outcome of our harvest, and attendant rituals are carefully observed to appease the deities and solicit their continuous goodwill. Although cassava is the major crop that provides the entire community with food, our ceremonial food crop is yams,which we celebrate in elaborate yearly ceremonies to honor Erot Edok, the yam fertility deity.

A folktale which has developed in the last ten years among Biase farmers reports that the cassava crop made its position known about the slight it had experienced from the human race. A woman who had gone to her farm one morning to harvest some cassava suddenly heard her cassava crop addressing her. The crop warned the human race that it would make itself difficult to get since it was not being accorded the respect it deserved, because yams, which barely fed the population, were accorded a higher status. If immediate steps were not taken to rectify this bias, the human race was going to find it difficult to feed itself. According to this story, farmers all over the world set aside a week in which they held festivities to honor and appease the cassava crop.

The Biase perceive the world beyond as being as concrete as the human world, but the earth is the nerve center from which man communicates with the ethereal world. Associated with this idea are associations and festivals which serve as channels of ritual sacrifices and religious observations meant to bring the goodwill of ancestral spirits and such earth deities as Erot Edok, Amretabi, and Efaoju for good harvests.

Later we shall see how these beliefs and institutions and other cultural factors have fared within the structural changes affecting rural communities. But first I give the founding history of the Agwagune of Biase, who are central to this book, as it was told to me by Onun Jonah Unoh Aquah Unoh Ikongo, who was told the story by his grandfather, who heard it from his mother.

> Today is the ninety-second year since I was born. I am Unoh Aquah Unoh Ikongo. My father was Aquah Unoh Ikongo. My mother was Ada Oje Unoh Ikongo. My parents share the same last name because they married from within the same house. The event that led to the founding of Agwagune was nothing more than a hunting expedition by two brothers, Ibara Edeget and Ibot Okom Omo Otutu, who came from the fifteenth-century Ekoi town of Agoi. When both brothers got to the present site of Ugep town, Ibot Okom decided to settle there, but Ibara Edeget moved on. He got to Abayongo, which was then uninhabited, and crossed to the eastern shores of the Cross River. Being a fisherman, he liked the place and put up a tent at a spot he named Uso Ekpa Eziba (the face of the earth) in Emomoro and lived there for four years. When he later went back to Agoi to collect his belongings, many people decided to follow him to his new home. To get to his home, the people were instructed in the native language to cross the river when they got to the shore, "Awa gwune" (Cross the river when you get there). The town of Ibara Edeget then became known as Agwagune. The first

person to rule Agwagune was Ibara Edeget, who founded the group; then Onun Eko Onungbe Inyang, who came from Esu Ikongo patrilineage; other earlier Inun were Onun Eko Effime, Onun Obagbo Egbai, and Onun Eko Oti. As my grandfather told me and said he was told by his own mother, it was in the reign of Onun Eko Oti that the Adim that are now troubling us came from across the river seeking refuge. It started one day with a flurry of messages and requests from some Inun across the river from us. They were Onun Obin Ukobin, Onun Ude Ayang, Onun Ebok Emarikongo, and Onun Obin Owari, all from Erei. They all sent messages to Onun Eko Oti, saying, "There are strangers on our coast looking for you. Please send people to come for them." Onun Eko Oti wondered briefly who these strangers were, then sent people to go and bring them. They brought the Adim, who explained that they were the Isobo, a small Igbo group driven from their land by a more powerful Igbo group. We sent a message to our brothers, the Ugep, who sent us five age sets in case there were any problems. We discussed where these strangers should be kept, and decided that they should be kept where they would serve as guards of the back entrance into Agwagune. The Onun decided to rehabilitate the strangers in his backyard and gave them the name, Adimi Mo Aba (those pressed to the ground) because they were driven from their previous Igbo region by a more powerful group.[4] The Adim reciprocated this kindness by paying an annual tribute of food, wine, and labor to Agwagune. The path in Emomoro village known as Ekat Ugom Adim (the path of the Adim) was the entrance designated for the Adim whenever they entered Agwagune to pay tribute. After the death of Onun Eko Oti, the Adim pleaded with the next Onun, Egbai Asang Inyang, to change their name, because Adimi Mo Aba reminded them of the humiliation they had suffered in war. The Onun then changed their name to Adim, granted them an independent status by marking out a formal boundary for them, and released them from paying further tribute. We also gave Abini its name, which in Agwagune is a form of

4. This may have happened in the same movement of the Igbo (reported by Ottenberg [1968] that occurred from about the 1700s to the later part of the nineteenth century when, in conjunction with the slave trade, there was a gradual movement of the Igbo from the Arochukwu and Ohafia areas to the south.

greeting, meaning, "You have arrived." We welcomed the Abini because they were a friendly group that sought economic and social relations with us.

Culture Change in Biase

Today Inun are less powerful than they were fifty years ago, and they are now more responsible to the government than to the people. This trend has been encouraged by the increasing testing and breaking of social rules and by the diminished power of traditional rulers as a result of their new status as paid government officials. Like other historically weaker ethnic groups, the Adim have taken advantage of this loss of status and have become more preoccupied with extending their boundaries than with fostering good relations.

Intercommunity disputes are frequent and have resulted in loss of life and property through offensive, defensive, or retaliatory raids and ambushes. Such disputes in Biase have centered around appropriation of resources and land use. The disputes have been between Agwagune and Akpet, Abini and Idomi, and Agwagune and Adim.[5] Intercom-

5. "Amidst protests, this man left his people to fight Odumugom-Agwagune in the hope that he was going to get himself two Agwagune heads for drinking water with. He came and met the very young men who had gone to mark our border. He pointed a gun and shot. The shot hit one of the boys on the left chest. The bullets tore tiny holes in his shirt but none of the bullets penetrated his body. The boy shouted, "He has shot me! Bullets have struck me but none pierced my body!" The boys started looking for him because he disappeared soon after he fired and emerged at another part of the bush. "Did he run away?" No, he was a man himself. He bent down quickly intending to disappear again but a shot rang out, 'Pai! Pai!' and the boys were already on him. They drew out their machete and separated his face from his head, then cut him right through the waist, and cut his head into very tiny pieces. That's when the trouble intensified. Our men were already very deep in the forest. Some started making their way to the trail. Their men started shooting at us inside the forest but their guns had no effect on our men. As we were passing through the forest, you would hear the gun right close to your head, you would hear the bullets whiz past, but nothing. But whenever we pointed a gun and shot at them, they fell and died. God was really with our men. Finally, our men all made their way to the trail. More of their men were lying in ambush just about ten feet away—more than two hundred of them were waiting to kill us all. They were on a strategic slope pointing guns down at our men. As our men appeared, they fired so much it was if the rain was pouring. Our men kept passing, with the bullets whizzing past them; the bullets kept falling on our men, 'tip, tip, tip'; but we kept passing through. However, if you had a problem in life or were not well prepared at home before the fight, you were picked out by the bullets and died there. But our men were passing through. People were wounded—about twenty or so—just wounds. Only four or six people fell and died there. They kept firing at our people from the bush and we returned the fire right at the point they were firing from. We really killed

munity warfare has resulted in the underexploitation of land because of the fear of encountering enemies who may be lying wait in the forest. Such disputes are difficult to settle because of the complexity of government involvement and the assertion of historical rights by warring factions.

Before the Biase came under the Nigerian polity, we had a political system based on male associations and patrilineages that regulated the social order firmly and punished offenders. For example, Biase communities such as the Agwagune and Erei are reported to have punished offenders by weighting them with stones and throwing them in the river. Frequently the instruments of such enforcement were the Abu male association and such mystical associations as Inyono and Ekpe; their authority was subsumed under the powers of an influential association Onun, Onun Abu or Onun Ikanda.

The younger Biase generation is seen as enjoying excessive freedom compared to how things were during the socialization of their parents and grandparents. There is now less respect for social etiquette and for elders and there is contempt for normative rules and irreverence for beliefs and practices seen by the older generation as essential to strengthening the Biase group.

For several years, we enjoyed political and economic parity with the Efik to the south and the Igbo groups west of the Biase, such as the Afikpo, the Arochukwu, and the Abiriba. These groups, who depended on coastal Biase communities to move their goods along the Cross River, have now eclipsed the Biase in Nigeria's political system. The Igbo were aided by the establishment of the seat of the government of eastern Nigeria in the Igbo region long before independence. Along with that came the building of schools and the provision of health facilities, the construction of roads to Igbo hinterlands, and the resulting incentives for local entrepreneurs to strike out on their own. The Efik coastal town of Calabar was the major port for European trade ships, and missionaries later established schools and trade centers in the area. Although the Biase resisted the intervention of European traders, they reluctantly accepted the Catholic missionaries many years later. However, when the colonial government was firmly established in Nigeria, the Biase found themselves outside of the system. Independence further isolated us as

them. It took their people three lorry loads to clear the bodies from the bush." Narrated by an Agwagune informant who participated in the fighting.

our neighbors belonging to the majority groups occupied government jobs at every level.

To imitate the new ways of life developing around us, a few Biase parents sent their children to secondary schools in Calabar, Arochukwu, and Afikpo. In my case, my grandmother sent me to Arochukwu to live with Dr. Alvan Ikoku, a Nigerian educator who was the proprietor of Aggrey Primary School and Aggrey Memorial College. It is common in southern Nigeria for parents to send their children to be raised by people they believe will teach the children good conduct and instill in them a sense of discipline and responsibility. My grandmother believed that Alvan Ikoku and his school would enable me to acquire those qualities.

As a member of a minority in Arochukwu, I learned the language and blended perfectly into the community. I always pretended to be a member of the Igbo group because I was ashamed to admit my minority status; I was helped by the fact that there was no one to speak Agwagune with. Whenever my grandmother visited me at school, I would steer her away from my friends so they would not hear us speaking in Agwagune. Toward the end of the school term, as my friends excitedly made plans to spend the holiday in Igbo cities, I tried not to be too forthcoming about where I was going.

By the time of the civil war seven years after independence, the Biase were solidly entrenched as a minority group in eastern Nigeria. Very few Nigerians even in eastern Nigeria knew our name while those who knew us referred to us in derogatory terms. We were not represented among the soldiers who made the decisions that affected us, and the handful of our people among the soldiers that fought the war never came back. The civil war significantly slowed down the effort of the Biase to educate their children. In the six years of independence, we flooded Igbo cities learning trades and unsuccessfully seeking employment in government offices. With the declaration of war, all of us abandoned our trades and schooling. For three years, the educational system in eastern Nigeria was stopped, and Biase students, like all students in eastern Nigeria, lost three years of schooling while the rest of the country prospered.

In that period the Biase experienced a rapid cultural reconfiguration as our traditional institutions were managed and controlled by strangers who had never heard of us or even knew how to pronounce our names. Soldiers made new laws that conflicted with our traditional behavior, decided which community members we were allowed to associate with, and upset our social and domestic lives. On many occasions soldiers would simply walk into our house, and without saying a word to anyone, go directly to our kitchen and take my grandmother's soup for their meal. For three years our villages were successively occupied by Biafran and Nigerian soldiers, none of whom came as friends or left friends when they departed.

THREE

Use of Environmental Resources

We started our court case against Adim on October 17, 1972. A small boy was assigned to assist me in my testimony during the case. Adim lost the case. Okoi Arikpo of Ugep was the lawyer for Adim. At the Federal High Court, Enugu, Justice Onyiuke asked how I knew the land belonged to my parents. I said the land belonged to them. I first addressed Arikpo, the Adim lawyer and said, "Arikpo, it was your father Arikpo Eget who brought you to this earth. He was the same age as myself. You and I are from the same ancestral parents. Do you want to give our land to Isobo people because they are paying you money?" All he could do was look down at the floor of the court. He couldn't answer me. Our lawyers were Otu Ekong and Orok Ita, who are both dead now. Adim had five lawyers. I asked them which tree had more power, one that is recently planted or one that sprouted on its own. My intention was to make them realize that the Adim, who were visitors to our area, knew little of its history and were fabricating everything they were telling the court. Adim lost the case. Three judges at the Federal High Court in Enugu gave us the verdict. Five hundred naira were awarded as cost against Adim and another fifty-four hundred naira as punitive damages. Until today the Adim have not paid that money.

(Onun Jonah Unoh Aqua, aged ninety-two,
the oldest man in Agwagune [April 1990]).

The need to enforce historical territorial rights is an important factor in the attachment rural villagers such as the Biase have to their land. As Ingold puts it, rural people "belong" to land "from which they are conceived to have issued and to have derived the substance of their being" (Ingold 1986: 140). The protection of that "substance," traditional beliefs and practices, as well as access to rights and privileges, is important in understanding how the Biase construct and sustain their territorial boundaries. In this chapter we will explore how human and nonhuman agents challenge the process of environmental management and limit the ability of some Biase villages to get along within their territorial boundaries.

Land and Territoriality

One of the more pressing environmental problems rural communities face is depletion of resources resulting from causes not directly related to the way the environment is used. Droughts and floods as well as territorial disputes and the encroachment of adjoining communities challenge the capabilities of rural villages to sustain or generate economic growth. The challenge for the Biase derives from the twin influences of flooding and politically intractable territorial disputes between groups attempting to balance the use of the environment and the management of intergroup relations. Both issues, one ecological and the other political, exert a significant influence on how this balance is achieved.

Until the late 1970s, Biase villages often cooperated in farming, the gathering of fruits, and hunting in each other's land on the basis of an informal short-term lease agreement. People from one village would simply go to another village and ask permission to use a piece of land. After breaking kola nuts and sharing palm wine, the visitors are shown land they may use for not more than one farming season. Under this arrangement, the Adim fished in the waters of the Agwagune; such cooperation also existed between Urugbam and Abanwan, between Okurike and Akpet, and between Ikun and Emomoro/Egbisim. In the past several years the Biase have, through marriage and trade, assimilated members of neighboring groups who have made their homes in Biase while retaining links with their natal homelands. Such assimilation has led to the development of a myriad of fictive kinship relations between people of distant Biase villages and has encouraged the development of relationships that are easily translatable into rights of mutual exploitation of resources. Groups such as the Agwagune consequently relaxed local rules governing economic cooperation with distant groups and extended resource use rights to these new community members. Such linkages also encouraged a few people in some villages to arbitrarily lease forest resources to friends and itinerant workers from neighboring villages. At the same time, community elders became inattentive to local sanctions that regulated such rights. This inevitably created some friction within some communities that were unprepared for such extensive cooperation. In 1989 Egbisim village was summoned before the Agwagune Inun Council for selling some of its trees to migrant lumber workers. The workers were told to leave the forest immediately and Egbisim was fined for breaching the rule. Other patrilineages have sold trees for lumber to people outside the community, and have sold palm trees to migrant wine tappers who sell palm wine within the community. In some cases the migrant forest users have encroached upon the rights of lineages not included in the original contract. When this happens, the workers are summoned by the lineage and charged a fee for their encroachment, or else they are told to leave the village.

Maintaining relations has not been easy, because children of immigrants may later assert what they perceive as their historical rights to resources. The agreements between individuals and groups about rights to resources are now characterized by frequent violation of the rules governing resource access and exploitation. Today opportunities for cooperative land use between and within communities have diminished. Warfare is increasingly used as a means of enforcing rights to resources; negotiation among villages who once saw themselves as bound by some form of kinship is now only sought for in state courts of law.

At the time of this study, Afifia and Emomoro villages of Abini had litigation pending in the state court over rights to farmland. Ordinarily such disputes are settled quickly through traditional legal processes. But such indigenous methods of conflict resolution have been superseded by the intervention of state laws.

In Agwagune, warfare was important earlier for maintaining territorial boundaries and for preventing strange groups from exploiting our resources. It is now less adaptive because the intervening role of the Nigerian state has forced the Agwagune to reassess traditional rules and locally sanctioned means of punishing adventurous neighbors. Onun Aquah's report at the beginning of this chapter suggests the frustration the Agwagune feel when dealing with the new challenges posed by the Adim in state courts of law. The increased authority of the central government has weakened traditional mechanisms for protection of property resulting in increased territorial encroachment by historically weaker communities. Without local mechanisms to protect the rights of the owners of resources "the management or self-management of resource use cannot be exercised any longer and, for all practical purposes, common property degenerates into open access" (Bromley and Cernea 1989: 17). According to Bromley and Cernea, open access "results from the absence—or the breakdown—of a management and authority system whose very purpose was to introduce and enforce a set of norms of behavior among participants with respect to the natural resource" (1989: 20). This breakdown has generated a series of conflicts between various Biase communities. Attoe (1990) refers to instances of warfare between groups attempting to increase their holdings of fertile land, fish ponds, and market centers. During my study there was a high frequency of conflicts involving land disputes between neighboring Biase groups. Hostility has reescalated in the last ten years between Adim and Agwagune, Akpet and Okurike, and Biakpan and Etono. At least five deaths were reported when warfare erupted between Abini and Idomi in the first half of 1991. Such conflicts intensify the vulnerability of these already-poor communities because exploitation of productive land is put on hold because of years of unresolved disputes.

Disputes are generally characterized more by the fear of the loss of civility within the group than by the engagement of feuding communities in hostilities. When there is an actual confrontation, the fear of harm the situation engenders among members of the feuding communities encourages maintaining distance as a strategy for individual and communal safety. In nearly all cases, feuding groups restrict their socioeconomic operations to territories well outside the conflict zone; they eliminate all forms of cooperation with each other; and in some cases, the villages relocate to more strategically defendable positions. It is apparent that territorial avoidance rather than conflict is the characteristic feature of Biase disputes. As the following chronology of territorial disputes among the Biase shows, not all conflicts result in warfare, particularly if parties in the conflict cooperate in other spheres or are not historically antagonistic toward each other. The frequency of these conflicts, however, underlines the difficulties of rural communities under the pressure to survive stressful conditions.

1. February 26, 1990: Emomoro summoned a meeting of its members following a report that Ugep was farming on Agwagune land sold to it by Adim. It was decided that a delegation should be sent to inform the Onun of Ugep that Agwagune viewed the matter seriously and would take any action to prevent it.

2. March 16, 1990: Five men from Okurike went to the bush to confirm a report that an Akpet man was cutting trees in their forest. They met the man and carried away all the trees he had cut. The man later reported to the police that he was beaten by the five men and that they had also threatened his life. He was able to identify two of the men to the police, who then went to Okurike and arrested them.

3. March 27, 1990: It was reported at a meeting of Emomoro/Egbisim that certain people were in their bush felling palm trees and brewing local gin from the palm sap. Such action is authorized only after a specific piece of land has been leased to brewers; on this occasion, the

lineage which owned the land had not leased it. On learning that Agwagune was going to take action against them, the men came to Odumugom and apologized for exceeding the boundary leased to them. They then made a fresh lease agreement with the offended lineage.

4. In April 1990, the Ase Egwa Imo lineage, which shares the Emomoro boundary with the village of Abayongo, met to consider a report that Abayongo had encroached upon and farmed their land. Ase Egwa sent two men to Abayongo to report the incident to the village Onun. The Abayongo Onun summoned the woman said to be responsible and asked her to withdraw from farming there immediately.

5. In July 1990, Abini learned that Idomi village had, in the company of government police, extended its boundary into Abini farming territory. There was great excitement in the village as preparations for war began.

Negotiating control of communal territory persists as an important aspect of rural resource management in Biase. In many cases production time is lost when hostilities erupt. In Adim, Agwagune, and Abini informants report that during conflicts, everyone stays away from the bush for fear of being killed by the enemy communities. For months before and after hostilities erupted between Agwagune and Adim in 1983 and 1986, farms were left unattended and bush paths grew thick with brush during the peak work period of December and January. Hostilities were not confined only to the principal villages of the conflict but also involved the Agwagune villages of Emomoro, Egbisim, Okurike and Itu-Agwagune, which on the basis of ancestral kinship relations, joined in planning and executing a raid on Adim.

Biase villages are continuously confronted with the possibility of losing their land not only to adventurous neighbors and governments but to other lineages within the same village. Where such infractions go unchecked, traditional land boundaries are obliterated; new borders constitute the basis for future territorial claims between lineages and villages. This often results in intralineage disputes; however, the consequences are often mild because of the close relationship between people in the same village. In a culture highly reliant on the knowledge of elders for the accurate interpretation of rules, of territorial rights, and of facts, incorrect assessments frequently occur, resulting in the loss of ownership. The following account illustrates a possible outcome.

In some Biase villages, members can claim ownership of resources such as trees by simply cutting a mark on the tree. The mark signals to others that the tree is taken and warns them to keep away. Often a relative or a friend is taken to the spot and shown the tree so the claim can later be corroborated. When the owner later needs to cut down the tree, he requests a meeting of the patrilineage, where he offers a bottle of locally brewed gin or palm wine and requests permission to fell the tree.

Ejim, a young man who was preparing to reroof his deceased father's house with wood from his father's tree, met his lineage to discuss the matter but faced opposition from another young man, Ogban, who said the tree belonged to his own dead father. Two elderly lineage members, aged sixty-six and seventy, supported Ogban's claim. However, a forty-four-year-old lineage member supported Ejim's claim, saying that he had gone with Ejim's father on many occasions to clear the brush surrounding the tree. He said Ogban was claiming the wrong tree and carefully described the part of the forest where Ogban would find the tree belonging to his late father. The matter was immediately resolved, and Ogban withdrew his claim.

On another occasion my patrilineage, Ase Egwa Imo, called a meeting of members to hear a report that one of our affinal relatives was farming on our land. When people dispatched to the site later returned to confirm the report, two elders were sent to inform the elders of the offending lineage. The lineage Onun called the person involved and cautioned him against the misdeed. The man protested that the land belonged to his family and that his father had previously farmed in that area. The lineage elders explained to him that a farm on that plot was leased to the man's father by a friend for one season's farming in 1964; his claim was subsequently withdrawn. Such misunderstandings arise when individuals lease farmlands to close friends whose children later assert ownership rights. Mistaken claims are not frequent, but when they do occur, there is always someone available to detect and rectify the mistake; if not, the new assessment becomes fact. However, if a mistake is not rectified this decreases confidence in community elders, whose judgments have traditionally been regarded as infallible. That younger members of the community sometimes excel their elders in certain domains of cultural competence (as in the case reported above) tends to further weaken the Biase authority system.

There is a possibility that the communal property system will degenerate into an open access system because of growing internal and external challenges to what is at best a weak system for the control of land rights. Bromley and Cernea make the important point that an important check against this tendency is an authority system that can enforce compliance. Among the Biase, there is a weakening of those structures that encourage sustainable patterns of resource use. The Nigerian state provides protection for external threats but is indifferent to internal conflicts, especially if the community in question is held in low esteem by both the dominant ethnic groups and the government (1989: 19). This strengthens Bromley and Cernea's argument that resource depletion in developing countries may have less to do with patterns of exploitation than with "the dissolution of local-level institutional arrangements whose very purpose was to give rise to resource use patterns that were sustainable" (1986: 7).

The Challenge of Environmental Hazards

While intercommunal warfare compounds the Biase environmental problem, flooding threatens our subsistence by decreasing the availability of land and riverine resources (such as trees, animals, and fish), by increasing the distance to resources, and by complicating access to external resources. Flooding is the most persistent environmental threat to our coastal communities. It is not a new phenomenon, but it was not until 1970 that flooding began to submerge residential lands. It is also possible that in the long term, farming may have some effect on flooding and landslides. Although much of Biase farming is on plots located about six kilometers from flood zones and on land that is relatively safe from flooding, cultivation may expose the soil to some erosion and to the transportation of sediments to low-lying lands. The accumulation of sediments over the years increases the potential for flooding by raising the flood plain and making floods more severe. The landslides occurring along Biase coastlines, however, are more likely to have been triggered by increased deforestation activities by upstream groups felling trees for lumber to meet the increasing demand because of the construction of more permanent houses.

In the past twenty years, the low-lying land on which Biase villages are located has weathered five disastrous floods which did extensive damage to houses and property. In 1989 the farm and residential lands of many villages were flooded, crops were destroyed, and there were massive landslides in the villages of Egbisim, Emomoro, Itu-Agwagune, Ugbem, and Etono. The main pathway that connects some of the villages was closed because of massive landslides. In Agwagune villages, buildings located close to the bank tumbled into the river. Agwagune informants report that in the last fifty years, such occurrences have resulted in the loss of 60 percent of the village land area to the river. The loss of houses is a major problem for rural people "because of the burden on limited finances in providing some replacement" (Cannon 1990). Although about 40 percent of the surrounding farmland is non-arable land, the soil holds a vast array of resources which the Biase use for fashioning tools, for house building, and for satisfying a variety of economic needs.

During the rainy season, about 30 percent of the farmland is under water. Many crops are submerged and farm plots become minimally productive. Biase farmers persist in farming these flooded plots because chances are the next flood may be less devastating than the one before it. In some years, their prediction is correct: the flood comes later in the year, making it possible for crops in low-lying lands to be harvested as soon as they mature. But in 1990, 20 percent of the cultivated land was already flooded by June, just two months after planting, prompting some farmers to start uprooting their crops to save what little they could before the cops rotted in the ground. Every farmer I spoke with mentioned losing between 60 and 70 percent of the expected harvest.

Until about thirty years ago, when the population of the five Biase groups was less than fifty thousand (table 1), such losses were buffered by the low person-to-arable-land ratio, less disaster from floods due to the availability of more choice land, and a greater tendency to cultivate long-term crops such as yams on land known to be safe from flooding. Today short-term crops such as maize, obiara, and cocoyams have become more economically important than yams and are commonly cultivated in floodplains more prone to flood disaster.

Although the Biase today number close to one hundred thousand, there are indications that we are not exploiting our environment with any more intensity than we did thirty years ago. First, our tools have remained unchanged, so the rate of intensive exploitation with such traditional implements as hoes, machetes, and stakes appears to have increased very little. Second, although the population of Agwagune has doubled in about thirty years, about 40 percent of us live away from our villages for at least six months of the year; of that number 90 percent of us do not own farmlands at home. So the effective Agwagune population is only a little more than the figure of 1963. Third, although Biase farmers now use wage labor, which suggests an intensification of farming, the wage labor essentially replaces family labor that is lost because of migration and the greater independence of younger family members (see chapter 6).

But these conditions have failed to stabilize the economy. In the last twenty years other factors have been more significant in the group's economy and have exerted greater influence on our managerial ability, thus making flooding much more oppressive than it was thirty years ago. The most significant disruptive factor is the escalating intercommunal hostility, which has resulted in underexploitation and wastage of arable land. It is possible that intercommunal conflicts are no more intense or even more frequent than they were earlier, but more arable lands were available for communal use then than now. Notwithstanding the Agwagune community's ownership of 70 percent of the land separating it from the neighboring Adim community, over 30 percent of this land has not been farmed for twenty years because of incessant warfare with the Adim. The land still being farmed is responding only minimally to the familiar flood-avoidance strategies because of the decreased accessibility of arable land.

According to an old cultural ecological opinion that war is likely to create safe zones for fauna to increase in numbers, the availability of forest land undisturbed for decades should be good news for Biase hunters (Ferguson 1989). But it is not. None of the people I spoke with found that game had increased because of warfare. Some hunters reported that it had been easier to find game in Biase forests in the years before the Nigerian civil war. Describing the ease with which he was able to find and kill game before the civil war of 1967–70, one hunter, Obazi, said, "Before you walk from here to Egbisim and back [30 minutes walk], I will already have two monkeys and a deer lying here for you. Today, you will find no monkeys for miles even if you sing for them with the voice of a bird." Faunal degradation may be attributable both to flooding and, indirectly, to the Nigerian Civil War. The five severe floods the Biase have suffered in the last twenty years have decreased the population of fauna; driven to seek shelter in residential areas, they are easily killed. A woman in Emomoro walked into her home and found a deer crouching in a corner of her kitchen during the 1989 flood. She quickly got help and killed it for food. Other animals may have found new habitats; for the few remaining animals the environment offers little protection. In the year I spent in Biase, the more common game animals brought home by hunters or caught in snares were pythons, crocodiles, wild hogs, and other animals which had better chances of surviving in a flooded environment.

A critical consequence of faunal migration is a decrease in the process of forest regeneration resulting from the absence of such seed carriers as monkeys, which are needed to disperse large fruits. The lack of regeneration is crucial in tropical forests where reforestation programs are absent and, as Myers notes, trees and plants depend on nonhuman seed dispersal agents (1984). Although the Agwagune do not appear to be overexploiting their forest, the lack of new growth jeopardizes the maintenance of a stable human-to-land ratio. Agwagune informants showed me a few remaining large fruit trees. Most of them have simply disappeared. Also relevant in this process of the loss of vegetation is the constant slow transformation of the forest even when a tree falls and destroys younger trees and plants and exposes the forest floor to light (Hecht and Cockburn 1990). Such processes are said to result in as much as a 25 percent tree loss within a period as short as fifteen years (1990: 30).

But the foundation for the exodus of fauna from Biase land may have been laid by the presence in Agwagune of a large number of soldiers fighting the Nigerian civil war between 1967 and 1970. Gunshots, mortar shells, troop maneuvers, and the general increase in the "hunter" population encouraged fauna to migrate to safer habitats. A further related factor, of course, is the clearing of bush by fire. But among the Biase this is not significant. Only small areas of bush usually burn, leaving substantial forest areas for new faunal habitat.

Understanding the role of faunal migration is important because the literature on coastal peoples such as the Biase repeatedly mentions hunting as a significant survival strategy (Basden 1966; Jennings and Oduah 1966; Harris 1965). Although there are still hunters in some villages, much of Biase land is now deficient in game because of human and natural causes. Hunting has lost its significance as a survival strategy.

Access to farms is a major difficulty during the rains. Routes are partly buried in mud and partly flooded, making the use of canoes necessary. Sometimes farmers are lucky enough to find a fisherman who will ferry them; most times they must wade through the water and mud. On a few occasions people trying to maneuver their way through treacherous farm routes have fallen and been seriously injured. Six accidents occurred in one week in August 1990; one involved the twenty-seven-year-old daughter of one of the chiefs, who fell and broke her collarbone when the load she was carrying fell on her. No one is reported to have suffered any known permanent injury as a result of farm route accidents, but Onun Obazi Esu, whose daughter was injured, said accidents are more frequent today because community members neglect to care for the farm routes. Until about ten years ago, the community would assign age sets to clear wood and debris from farm routes before the floods set in and to clear off brush after flooding. This made traveling less dangerous for farmers in rainy and dry seasons. Now less attention is paid to these tasks of community interest. Elders like Onun Esu blame the young for this neglect of the community. "When I announce work for the age sets, only a few children turn up because the parents tolerate the disobedience of those who do not. If children refuse to work, will an old man like me do it for them?"

The young in Emomoro respond that their elders lack leadership skills, which prompted them to form an association of five younger age sets to organize and execute communal tasks they believe are often neglected (see chapter 6). Obazi Edet, member of the Congo age set, who heads the association said: "Our elders have failed us. There is no project you leave for them that is not bogged down by arguments. That's why we the younger age sets formed this association."

Although people attribute the worsening of access routes to communal neglect, the routes are also deteriorating because the presence of more people in the villages results in higher use. The population of the five Agwagune villages (table 1), for example, increased by 61 percent between 1963 and 1983 (from 4,752 to 7,784), and by the time of my research in 1990 it was estimated to be 9,256, about double the 1963 figure (Cross River and Akwa Ibom State Population Bulletin 1983–90).[1] This reason is, however, not significant because a large number of people live outside the villages.

Flooding also increases the risk of disease through contamination of drinking water and through accidental death by drowning in swollen rivers. When the Cross River and its tributaries begin to flood, the clear river water we drink turns brown and murky from mud and upstream debris. Informants report that in 1989 they saw two decomposed corpses at different periods floating downstream into the sections from which we drink. Such sightings are common. People frequently drown when using small dugout canoes in the dangerously swollen waters.

As much as possible, we avoid drinking river water during the rainy season; instead we put out buckets and basins to collect rainwater

1. The 1963 census figures are not reliable because the figures were widely disputed. Since about 40 percent of the Odumugom population lives outside the community, pressure for land is almost nonexistent, because the resident population of just under 6,000 is only a little more than it was 30 years ago; however, the age range of independent adults is also lower now than it was 30 years ago, resulting in a possible increase in the volume of traffic along the farm routes as more people are now using the routes. Thus the increased rate of use is likely to worsen wear during floods.

from our roofs. During my stay in Biase it rained at least twice weekly in the flood period of May to August. After about the end of August, the rains ceased temporarily, marking what we refer to as the "August break"; the buckets were empty and people had to resort to drinking water from the river. It is difficult to measure the health consequences of a flood since, as Cannon suggests, the threat to health may continue for years (1990).

Reports from various regions of the world mention the positive effects of flooding in improving soil quality (Cannon 1990); the Biase are no exception. Despite the ecological problems from flooding, patches of land exist where thin alluvial deposits permit the growing of vegetables. Shortly before the end of October, the river begins to recede and sand appears in section of the previously swollen river, gradually spreading over the entire coastal breadth of Emomoro, Egbisim, Ikun, Okurike, and areas much further down the coastline (plate 2). This extensive sand bed might suggest a potential improvement in the quality of the soil if it were not totally covered with infertile sand except for a 5000-square-yard tract that supports the cultivation of okra for a few weeks. A greater economic potential of this sand lies in its value for making cement blocks. Unfortunately, little can be done with the sand because of the difficulty of moving it from the coast to the cities where it is in greater demand.

The primary difference between coastal Biase villages and the non-coastal villages, such as Abini, is that the latter have a relatively more habitable environment. Abini is situated on a firm slope having a mixture of sand and stone that provides a good surface for runoff water; there are rarely problems of flooding. When the Ubam and Udip streams dry up after the rains, people in Abini drink water from ponds; this results in infections and deaths from various waterborne diseases. However, during my study there were no reports of death attributed to poor drinking water, probably because other culturally assigned causes readily provided alternative explanations.

Environmental challenges to human survival capacity have always existed. Rural peoples have often confronted it appropriately. The Biase are not now experiencing environmental challenges that were absent, say, two hundred years ago. In recent times, however, these challenges have become more serious because of a combination of factors that now overwhelm coping mechanisms that were adequate a few decades ago. For instance, a shortage of labor, increasing distance to farms, and less friendly neighbors all limit the ability of farm families to respond to disasters.

Conclusion

We must understand the encroachment upon rural territory by neighbors whose subsistence strategies have become equally dysfunctional in order to understand the dynamics between the Biase and their environment. Increased environmental problems and the need to ensure sustenance for present and future generations escalate territorial conflicts, intensify people's desire to hold on to historical notions of the ownership of resources and prompt neighboring communities to assert rights to territories from which they have been historically excluded. The increasing warfare between the villages hurts the rural economy by curtailing access to resources. Despite conflicts over access to resources, the Biase are not going hungry. We still produce enough food to meet our subsistence needs. In the next chapter I examine the influence of Biase indigenous technology on resource management and the response to environmental stress.

1. The swollen Cross River in August

2. The Biase section of the Cross River dries up in October leaving an extensive sand bed.

3. Many Biase villages are situated within mangrove swamp forests.

4. During the rainy season, waterlogged, mud-filled routes
are the only access to farms.

5. Bamboo bridges help farmers maneuver the difficult farm routes in the rainy season.

6. Plantains are often cultivated in home gardens.

7. Lineage members assemble for farmland allocation.

8. A blacksmith fashioning blades for hoes.

9. Yams are associated with men's farming but are occasionally cultivated by women.

10. Cassava is intercropped with yams by cutting old stalks and burying them in the earth.

11. Maize, cassava, and obiara growing in yam mounds.

12. A farm with stakes ready for yam shoots to climb.

13. Men know how to design wooden handles for hoes.

14. Men who own large yam barns are considered good farmers.

15. Pond fishing requires many people to wade through the trapped water with hoop nets.

16. Fish is shared among lineage members after pond fishing.

17. Cassava skin being peeled off in preparation for gari processing.

18. A fishing trip down the Cross River yields a much-appreciated food.

19. Members of the Aka-e-Mitin Aneba celebrate the end of a successful meeting.

20. Some members of the Guinea age set.

21. Men spend their leisure time in a lineage square.

22. Women are required to be in the market on market days even if they have nothing to sell.

23. Thatch roofing material being fashioned from rafia palm leaves.

24. Corrugated iron sheets are increasingly replacing traditional thatch roofs.

25. Agwagune village heads and lineage elders at a meeting of the Inun Council.

26. A lineage head is not necessarily the oldest member of the lineage.

27. A libation being offered by a family member.

28. Erot Edok maiden dancing during the New Yam festival.

FOUR

Managing the Environment

> You spend all that money to come from America just to find out what I do on my farm? Well what I do is make food for my family. Sometimes food just sits on my farm and dies because I have no place to keep them. Sometimes, I just give it to people in my family. I know that if you ask them they'll say I never give them anything. Instead of asking me all these questions, go and ask those people that sent you how Egypt was able to stock up and store corn over three thousand years ago through seven years of famine.
>
> (An informant in Agwagune, Biase.)

Although I earlier suggested that Biase technology has changed very little over the last several years I do not mean that no aspect of our means of subsistence has been modified. Continued relationships with Afikpo fishermen and migrant laborers and traders and even the presence of government export-crop farms, have in some ways modified our technology: machetes now come with strong plastic handles, although there are still machetes with wooden handles; there are new crop varieties alongside the old ones; our hoe now has a curved wooden handle and a wider blade made of durable steel with its diameter increased from 6 inches to 10 inches. With this new hoe, our farmers now construct bigger earth mounds averaging 30 inches high and 48 inches wide. Some of our farmers now build 10 to 15 more mounds than when they used the smaller hoe. Farmers in the middle belt of Nigeria who use hoes with this large blade report that the hoes make their work faster by increasing "the amount of soil scooped up with each digging" (Ajayi et al. 1990). While this new hoe makes our farmwork easier, the plots we cultivate have not increased beyond the average of 2 acres divided into 3 or 4 plots totaling some 2,880 square yards. (0.60 acre or 0.240 hectares). But perennial flooding of farmlands has taught us to counter environmental challenges by planning our farming activities in anticipation of the July and September floods and to reduce the impact of disasters by matching crop to soil type. We cultivate multiple plots to spread the risk of flooding, poor soil quality, and rampaging hogs; we adopt high-yielding crop varieties; and we increase the diversity of crops and minimize our dependency on staples.

Soil Taxonomy

Knowledge of soil types is particularly important in years of high flood, when lands that support certain crops at certain times are covered with water and mud, thus challenging the effectiveness of local land management. Since there is no way the Biase can predict the severity of flooding in any particular year, land use decisions are conditioned by the possibility of disaster. The Abini and Agwagune classify their soil into six categories: *idep, ogum ebe, ebe, ebem, obum, and ikwu.* Ebe is loamy and is favored for intercropping corn (plate 11), cassava, and cocoyams with yams; plantains and some vegetables are also common on the land. Ebem is a flood plain and usually floods very early. It favors fast-maturing crops such as maize, which can be harvested before the floods set in, although yams are also planted on ebem land. Ogum ebe is mostly used for yam cultivation, but yams actually do better in idep than in other types of soil. Farmers say the loamy quality of idep keeps yams in the mounds longer even during the intense heat of the planting weeks, thus permitting normal growth. Obum is sandy and loose but good for growing corn, vegetables, and cassava. Ikwu, land used for the home garden, is distinguished by thin alluvial deposits from the receding floodwater; it is used by women for growing vegetables and is usually cultivated in November, two months before the general farming activities begin. It is also the first area to be flooded.

Some farmers say they do not invest their resources in certain land if they expect flooding in a certain year. Sometimes the appropriateness of a particular soil may not be a strong reason for using it. For example, despite general agreement by Abini informants that yams do better in idep than in ebem soil, Abini yam farmers planted more yams in ebem than in idep. One reason was the greater possibility for multicropping in ebem. Ebem mounds are, however, more susceptible to erosion because of the soft texture of the soil. Farmers therefore make ebem mounds larger than idep mounds so women can plant a variety of other crops alongside the yams, to hold the soil. Ebem is usually closer to the residences (3 kilometers as opposed to 6 kilometers for Idep), shortening the distance for transporting yams from the farm. To some extent, these soil types affect the amount of land some people cultivate. Older farmers tend to own fewer plots and farm close to the village to avoid walking long distances. Within this range, the land available is ebem, which has advantages for multiple cropping. Generally land distribution is not affected by soil types because every farmer has an equal right of access to the land in each farming zone during land distribution. A farmer knows how to match specific crops with the different soil types, and makes her cropping decision on the basis of the crop she wants to produce rather than on the basis of soil quality.

Differential Management of Farmland

Because different types of soil support different crops, farmers practice differential farmland management strategies to distribute their chances of having a good harvest over fields of several different soil types. This tendency, referred to as "fragmentation" by some writers, is sometimes considered a major problem hindering rural agricultural development,[1] "the blackest of evils, to be prevented by legislative action as one would attempt to prevent prostitution or blackmail" (Farmer 1960, cited in Bentley 1987: 31). To correct this evil, our farmers should consolidate their farmlands. Considering their native knowledge of the ecological challenges in many rural ecosystems, it is surprising that even some African scholars (for instance Nwankwo 1981, cited in Kiwanuka 1986) fail to articulate the disadvantages of consolidation or of massive mechanization of rural farms in some African ecosystems. Often land consolidation advocated by "expert" management is less adaptive because of tough terrain and greater susceptibility to flooding.

There are other economic considerations. Although land consolidation has been linked with an increase in cash crop production, particularly in industrialized Europe, it is uncertain that the amount of money expended on those programs was economically justified (Bentley 1987), nor have such experiments succeeded in parts of Africa. For example, when the government of Kenya mandated land consolidation the number of personnel involved for some communities was staggering. Bentley reports that in the Kiambu location alone, "about 420,000 plots were measured and consolidated into 50,000 holdings by 98 committees with 2,750 members, a staff of 14 senior officers, 500 surveyors, and 1000 laborers" (1987: 60). Reporting later on the results of this gigantic experiment, Green noted that consolidation has marginalized

1. Fragmentation, of course, implies a partitioning or segmenting of a homogeneous entity, which in reference to farmers means that they are fragmenting a range of homogeneous "good," or farmable, land.

about 500,000 people in Kenya, who are now unable to spread production risks because they are limited to cultivating poor land with submarginal rainfall and a high risk of drought (Livingstone, cited in Green 1989). Plot dispersal is not really a question of right versus wrong but of what works best under a given set of environmental conditions. The Biase, like many farmers that practice differential farmland management, are simply making the best decision under the existing social and ecological circumstances.

Multiple plots are beneficial not only because they afford people the opportunity to produce their own food and fulfill their obligations to kin; they also serve to reserve certain fields for less capable community members and extend socioeconomic relations beyond the immediate environment to neighboring communities. Among the Biase, Urugbam village has leased land to Abanwan villagers who are short of farmland; Ikun has leased land to Egbisim farmers; and Egbisim has leased land to Abini. The land leased is often part of three or more dispersed plots the farmer needs to sustain his family. Some Biase lineages who do not own much land exercise greater control over the use rights of members. Thus, considerations of individual welfare often are a strong reason for multiple plots among Agwagune farmers. Patrilineages with excess land may lease portions to other lineages. For example, the Ase Egwa Imo patrilineage of Emomoro offered land to two lineages that were prevented from farming their own fields because of a 1970 flood. The leased lands reverted to Ase Egwa at the end of the farming season.

Dispersal of plots is also affected by marriage. Agwagune women who marry outside their lineages retain land use rights in their matrilineal descent groups if they have kept up their moral and financial commitments to the group. Such community members own farmlands in lineage lands that are located in different and far-flung parts of the villages. When the Igbadara patrilineage began to distribute farmland, someone said that one woman who already owned four plots from her patrilineal and matrilineal kin groups should not be given a plot from her husbands's patrilineage because she would be unable to handle the work. This provoked angry retorts from people on both sides of the issue. The woman was later given a plot when elderly lineage members said that she had never been known to abandon her farm plots, and that she was on good terms with her husband's patrilineage because she did not owe any money. Igbadara had to give the woman a plot so as not to antagonize her lineages and jeopardize future cooperation.

Another reason for dispersing plots in Biase is that it is customary to give choice plots to women who have recently lost their husbands and might be having difficulty in finding labor. According to the lineage Onun of Ase Egwa, "When we go to 'break the earth' [plate 7], we look around for land to give to women who are without husbands; they are called 'sorrowful ladies.' After the others have gone and have finished 'breaking the earth' and taking their own farmlands, we take these women to land that requires little work but is good for cassava and vegetables, and apportion their own farmlands to them."

The multiple-plot strategy has had a significant effect on the adoption of fast-maturing, high-yielding crops because of the possibility of year round cultivation offered by different soil characteristics. Crops planted in floodplains are harvested before flooding; late planting in higher areas allows for a later harvest.

In spite of these advantages, multiple-plot farming also presents the Biase with some problems. While the multiple-plot strategy seems to be the best model for Biase farmers, it faces the new problem of a lack of sufficient labor to maximize its potential. A common complaint during clearing of bush and tilling of fields is that there is no labor to hire or money to hire labor. Farmers are often unable to find the time or money for the labor required. In a few cases the enthusiasm with which some people take on additional farmland often is not equaled by their readiness and capability to do the work needed to develop it. Wild pigs sometimes pass through unused land into developed farms, destroying crops. Some lineages attempt to help the situation with firm rules punishing people who leave their plots uncleared. In the year of my research, some lineages imposed a fine of one hundred naira on people who left their farmlands undeveloped. This rule tends to check land waste because farmers only ask for the amount of land they can comfortably cultivate.

Some farmers, unable to develop their farmlands because of insufficient labor, get other farmers to help them. But where sufficient labor is lacking, farmers fend for themselves and do all the necessary work on their farms. Some farmers find themselves short of time because of other economic responsibilities. One farmer noted that she had been expecting her sons to visit her during the last planting season but that they failed to come because of their commitments in the city. As a result, she was unable to tend some of her farmland adequately and the neighboring farmers were angry because their crops were damaged by wild pigs. On another occasion, a farmer had arranged to obtain extra labor from the young men in the community but was disappointed when they did not turn up because they were dissatisfied with the wage rate.

When farmlands are left undeveloped, damage to adjacent farms by hogs is usually resented even by sympathetic farmers. A farmer took his brother before the patrilineage Inun for failing to develop a plot. To strengthen his case, he cited an incident when hogs passed through an undeveloped plot and attacked and killed an elderly woman working on her farm. Such cases are viewed seriously by community members, who impose stiff penalties on offenders. However, since the existence of undeveloped plots does not mean that a farmer is not cultivating other plots, the problem is insignificant and contributes little to land wastage in general.

Persistence of a shortage of labor results in low production per unit of land. Such a decrease in production would not be a problem if the resident rural population were the only consumers, but Biase immigrants to the cities regularly return to their villages to get supplementary food. Outmigration from Biase (see chapter 5) has not had a positive effect on the group because off-farm employment is difficult to find, particularly for the largely unskilled Biase immigrants. Many who find jobs in the city occupy low-level positions that barely pay the rent. The shortage of labor affects the benefits of multiple plots by limiting the population's capacity to minimize its risk: fewer multiple plots are cultivated even though it seems to be the best strategy.

High-Yielding Crop: Obiara (Cocoyams)

The new high-yielding obiara has restructured food production among the Biase by challenging the gender-based ideology on cropping and by competing with perennial staples such as cassava. Obiara is a variety of cocoyam that Abini and Agwagune informants say was introduced by Afikpo settlers about twenty years ago. Until 1980, little interest was shown in this crop because an older variety of cocoyam was still common. But the older variety has been replaced by obiara, which is eaten daily by all households in my study. It serves as an important supplement to cassava and is cultivated by both men and women.

Considering the cultural importance of yams in eastern Nigeria, it is surprising that obiara, which was uncommon among the Biase as recently as the early 1980s, was generally ranked higher than yams both as a staple (fig. 5) and as a cash crop in 1990 (fig. 6). In a survey of thirty households in Emomoro/Egbisim, obiara, cassava, yams, and maize are the crops most cultivated and constitute an important part of the daily menu (fig. 5). Some farmers cultivate plantains and other marginal cultigens, but these were rarely mentioned because they are classified as

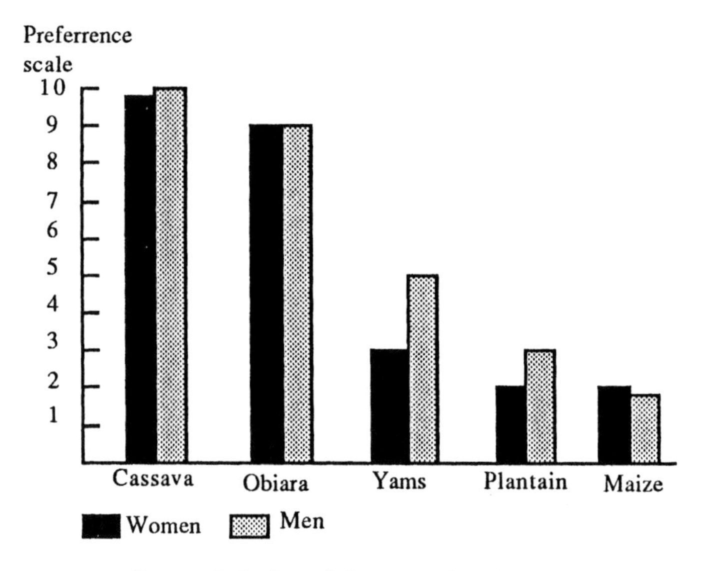

FIGURE 5. Preferred Consumption Crops

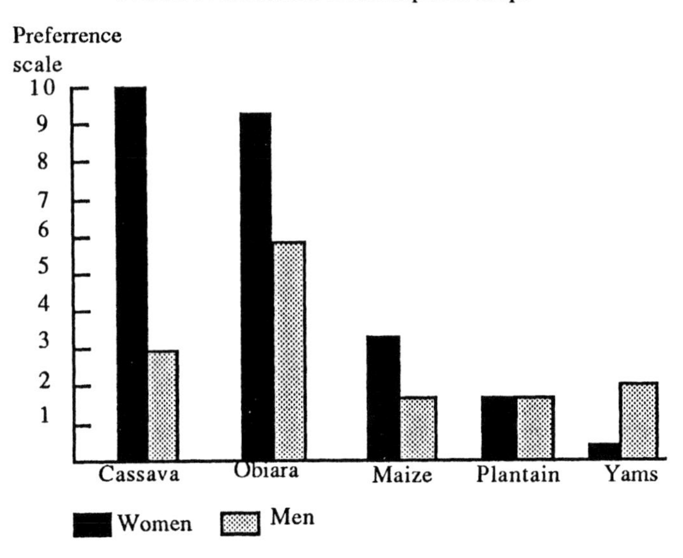

FIGURE 6. Preferred Cash Crops

supplementary foods ("something to eat in addition to my cassava"). Obiara is said to be high-yielding, matures quickly, demands less work than yams, and has a more ready market than yams (table 4). It is more delicious than yams, and because it is high-yielding, a household can sell a substantial quantity and still have enough left for food. Obiara was

TABLE 4 Reasons for Adopting a New Crop: Obiara

	High (%)	Medium (%)	Low (%)
High yield	16	36	48
Early maturity	30	44	26
Less work	0	13	87
Ready market	87	13	0
Greater demand	94	6	0
Better taste	80	20	0

Note: $N = 15$.

ranked just a little below cassava (8 on a 1–10 scale) as the most preferred consumption crop in Biase. Yams were rated third on the scale by 85 percent of the Abini sample. Five households that rated yams above obiara strongly disagreed with the suggestion that obiara is more profitable than yams. They said you could sell a string of yams (between 30 and 40 medium-sized tubers) harvested from about 15 to 20 mounds for about 300 naira, but you needed about an acre of obiara to equal that amount from selling obiara. Obiara has a production cost comparable to that of cassava, a much lower cost than yams,[2] and is ranked almost as high as cassava as an answer to my question asking what crop most farmers prefer to cultivate.

The minimal financial consideration in cultivating cassava partly accounts for why we cultivate cassava more than any other crop. The cassava farmer either borrows or reuses old cassava stalks, which are cut into 4 to 6 inch lengths and buried in a new plot to reproduce a fresh crop of cassava tubers (plate 10). In addition, the financial resources of our farmers are so meager that a crop requiring even a relatively small financial input will be given less consideration than one that requires little or nothing. In my survey, 61 percent of farmers said they would invest more resources in cultivating cassava than in other crops, 52 per-

2. I interviewed two yam farmers and two obiara farmers to get information on what they spent in cultivating both crops over a period of nine months. Since some farmers had planted yams before I began my research in February, I counted the number of yam and obiara mounds to estimate the quantity of seedlings and calculate the cost of yam and obiara seedlings based on the current prices for the 1990 farming season. I classified all labor (weeding, cutting stakes, guiding the climbing shoots, harvesting, constructing barns, conveying yams from the farm to the barns, conveying yams or obiara from the barn or farm to the village or market) expended on each farm as paid labor and calculated the labor cost on the basis of the going fee of about seven Naira per day. The yam farmers spent about eight hundred Naira buying yam seedlings, and an average of 182 Naira in hiring labor. Obiara farmers spent about thirty Naira on obiara seedlings and sixty-three Naira on labor.

TABLE 5 Food Storage

Food	Place	Duration[a]
Cassava	Soil	3 years
Gari[b]	Bowl, bag	4 months
Obiara	Soil	3 months
Yams	Barn	6 months
Maize (dried)	Cooking area	12 months

[a]Possible.
[b]Processed cassava.

cent picked obiara, and 10 percent picked yams. This seemed a bit odd to me, since income from obiara is more regular than income from cassava, and obiara requires less capital input than yams. It is tempting to assume that the need to satisfy domestic requirements rather than to make a profit may be the reason why cassava was chosen over obiara (fig. 6). But we shall see later that cash considerations are also a factor here because gari, obtained by processing cassava, has a higher demand than Obiara, particularly in the cities. Although obiara brought income faster, it was wanted only by itinerant Afikpo traders, and farmers felt a pressure to sell obiara quickly and cheaply so it would not rot. On the other hand, gari could be stored a little longer until prices were more favorable (table 5).

Besides storage problems obiara cultivation also faces the problem of a lack of adequate transportation. Obiara harvesting begins at about the peak flooding season of July (fig. 7), when farmers harvest it so it

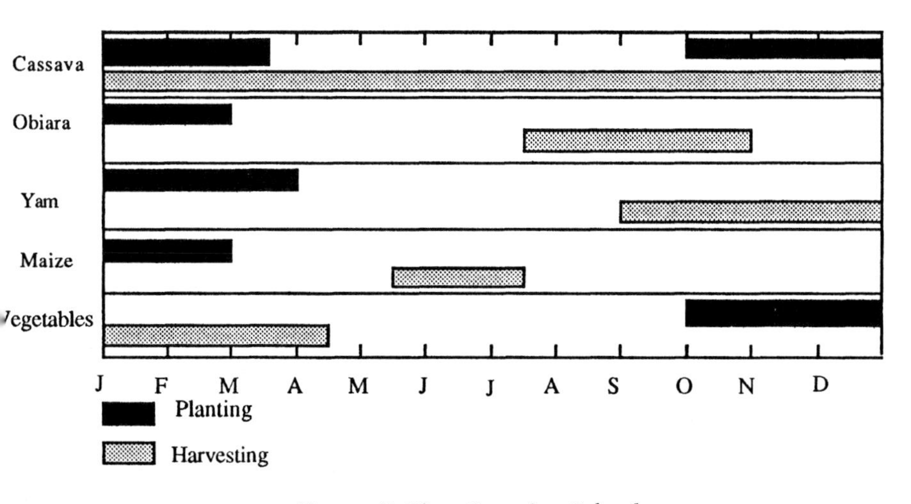

FIGURE 7. Biase Cropping Calendar

TABLE 6 Means of Transportation Owned

Village	Canoe		Bicycle		Motorbike	
	Men (%)	Women (%)	Men (%)	Women (%)	Men (%)	Women (%)
Emomoro	85	6	2	0	0	0
Egbisim	75	3	1	0	0	0
Abini	40	1	20	2	2	0

Note: $N = 87$ women, 117 men.

will not rot in the water-logged soil. This is also the time when vehicles stop going to the villages, so the difficulty with transporting obiara quickly to outside market results in some loss from rotting. Canoes and carrying on the head are the customary means of transporting goods to neighboring markets in the rainy season, but only a few women who sell this commodity own canoes (table 6) that are large enough to be safely maneuvered in the swollen Cross River (plate 1). In this regard, the cassava crop has the advantage of keeping longer in the soil or in its processed form, so the farmer has greater control over its sale and consumption.

Contrasting Valuations of Two Major Crops

Although caṣsava has always been the main food crop of the Biase, it has never had the prestige of yams, partly because it is defined as a woman's crop. I noted that while cassava is the main staple of the Biase, it is yams that are celebrated in elaborate yearly ceremonies.

This contrasting valuation of the two crops has been examined in terms of Malinowski's notions of how the Trobriand Islanders used magic to counter the hazard imminent in sailing (Brosius 1988). Since Trobrianders are unable to control the rough seas, they seek comfort in magical rituals meant to ensure safety. Brosius reports that among the Ifugao of the Philippines the sweet potato is the most important of all food crops, but rice is the crop that is highly valued and that defines important features of Ifugao society. The Ifugao cultivate about 30 percent more sweet potatoes than rice, and in some districts sweet potatoes equal or surpass rice in importance. But the Ifugao despise sweet potatoes and feel insulted if they are said to have only sweet potatoes to eat (Barton 1922, cited in Brosius 1988). The explanation is that rice is much more difficult to cultivate than sweet potatoes, because it is easily damaged by pests and other environmental hazards. The wish to reduce those risks and uncertainties is partly responsible for the much greater ritualization of rice than sweet potatoes, which are more common (Brosius 1988). This means that since cassava occurs more widely and is less uncertain and less expensive than yams to cultivate, the Biase have little need to resort to magical rituals to ensure its safety. But this notion does not adequately explain the Biase situation, because cassava is no less at risk than yams. Both crops are subject to the same environmental hazards, particularly flooding, which the Biase farmers face annually.

Brosius also refers to a labor theory that assumes that the successful production of a crop which requires a great deal of labor may result in the high valuation of that crop. Cultivating yams entails building the mounds, buying yam seedlings, cutting stakes for the climbing shoots, and consistent weeding. Cassava cultivation entails less effort. The farmer merely cuts old stalks and buries them in the earth. The greater effort expended in cultivating yams may thus account for their higher valuation. While less labor is expended in cassava cultivation, more income is obtained per unit of cassava than for yams which sell for a higher price but are less in demand than cassava.

The higher valuation of yams may have a historical explanation. Biase men and women named yams as the most respected food crop. It is respected because yams used to be the major commodity of long-distance traders in southern Nigeria. Since men dominated this trade, they also tended to control the production of the crop. Yams are thus associated with men's farming in much of southern Nigeria (Martin 1988; Uzozie 1981; plate 9) and were historically important in making Agwagune men wealthy when they traded up and down the Cross River. Yams still have the status of the prestige crop of men in much of southern Nigeria, giving men political and economic dignity (Basden 1966; Forde 1964; Uzozie 1981).

With that background, I found it culturally intriguing to notice a high participation by men in cultivating obiara. Since obiara are a variety of cocoyams, it is usually considered undignified for men to cultivate them. For the same reason, it was significant that 22 percent of male farmers cultivated cassava although they still perceived it as a female crop.

Such changes in gender relations to crops have occurred in other cultures when a new technology or a price incentive resulted in men joining or assuming control of crops traditionally classified as women's (Gladwin and McMillan 1989; von Braun and Webb 1989). For example, in The Gambia the upgrading of rice-production technologies which was meant to improve the welfare of women who were traditional rice farmers, had the reverse effect of putting control in the hands of men, who were attracted by the labor-saving factor of the new technologies (von Braun and Webb 1989; see also Carney and Watts 1990). While a price incentive was also a factor in the Biase case, it did not result in men assuming control of either cassava or obiara. But it was attractive enough to persuade them to negotiate their socially recognized role. However, women tended to have two or more obiara plots compared to just one of men. No one during my study named yams as a major source of food, and only a few men admitted that they were a significant source of income. Yet its historical and symbolic appeal are very strong and men who own large yam barns (plate 14) are still regarded as good farmers, not only in Biase but also among other southern Nigerian groups (Uzozie 1981).

Biase farmers see the comparatively low labor requirements and financial cost of obiara production as an important incentive for raising obiara to compensate for their loss of income because of the low profits from yams. Obiara are particularly helpful for women, who have found a new source of income at little cost, and obiara production is also appreciated by men experiencing lower profits from their "male crop" who are willing to suspend cultural codes to improve their chances of survival.

One may wonder why rice is not an important crop here, since there are wet conditions for part of the year. Although some farming groups in eastern Nigeria cultivate rice, this crop is not important in much of Biase because the farming history of the group is built around crops that are much more in demand within the marketing zone of the neighboring villages. Rice was for a long time regarded as a ceremonial food, eaten only on Sundays in many homes or in celebration of such alien festivities as Christmas; or it is offered as a delicacy to an esteemed guest. It was therefore socially and economically less important than such staples as cassava, yams, maize, and plantains, which the Biase were technologically better equipped to produce in substantial quantities. The difficulty with transporting foodstuffs to outside markets is also cited as a reason for the lack of interest in rice. Rice is heavier than other food items, which the Biase mostly carry on their heads to neighboring markets. Fishing is the primary occupation of Biase men and women during the rainy season. It is a more certain source of income than rice and is easier to manage alongside the farmwork demanded by perennial staples.

While most Biase communities fish during the rainy season, the

Abini, who do less fishing than other Biase communities, cultivate rice during the wet season. Indeed, just as the Agwagune rely on their fishing for much of their rainy season income, it is rice cultivation that Abini informants say brings them the money they will invest in next season's cultivation of yams, plantains, and obiara. Rice is also a ceremonial food among the Abini, but it is grown primarily as a cash crop. Some farmers reported that in hazard-free years, they harvest up to thirty bags of rice and sell them at about three hundred naira per bag. Young Abini farmers suddenly find themselves with an enormous amount of money and spend it as quickly as it comes in. But Biase farmers rarely experience a hazard-free year. Rice farmers had a bad season in 1989 when many coastal towns of Cross River State experienced an unusually bad flood. Most farmers lost all the rice they planted. The Onun of Abini said he invested about two thousand naira in rice farming but did not harvest a single bag. Despite the bad season, he doubled his investment in rice. It is often suggested that such misfortunes make the small-scale farmer reluctant to take risks in the future. But the rice farmers I spoke with said that it was merely a bad year and the next year would be better. The experience was not enough to make them abandon rice farming, although they had lost much money in the disaster. Later investigation showed that about 60 percent of Abini farmers who cultivated rice in 1989 were unable to continue in the year of my study. Many had taken government loans, which they were unable to pay back, and they were struggling to meet a twelve-month extended deadline for repayment.

Generally, the Biase technological strategies have been undermined by factors which the group can no longer control. Knowledge of the soil is now less important because of the alternative uses the land has to serve; multiple-plot ownership is less beneficial because some plots are in unsafe locations; while adoption of the high-yielding obiara has increased farmers earnings, their high perishability and the resultant need to sell them quickly transfers much-needed income out of the villages. So the financial gains from a greater yield are short-lived because of the problems of storage and high perishability. In the following sections, I suggest that farmers are producing surplus food but that the gains are not sustainable because of seasonality, overconsumption at the production bases, the lack of means to preserve food for longer than a few days or weeks, and the inability to get food to distribution points fast enough.

Food Processing and Storage Technology

The lack of adequate storage facilities is one of the major reasons for the food crisis in Nigeria. The statement at the beginning of this chapter was made by a bemused farmer who asked me why there should be a storage problem when Egypt, with a relatively limited technology, was able to stock up and store corn over three thousand years ago through seven years of famine. If the answer to this question were readily available, the food crisis in Africa would not be as intractable as it appears to be at present. But the answer is not available. As a result, a significant amount of food is being wasted through needless consumption, sale below cost, uneven redistribution, and high perishability. Ironically, when governments and developers plan for surplus production among small-time farmers, they pay little attention to how the projected surplus will be stored because it is easy to forget that farmers are already producing a surplus. Since it is estimated that about 30 percent of rural production is going to waste,[3] we should wonder how people could survive such adversity if they were not producing a surplus. When the farmer plans for the season's farming, his plans for plot size take into consideration expected significant postharvest losses (table 7) from such unpredictable factors as unfavorable soil conditions, perishability, bad market conditions; and the addition of a member to the household or the loss of one.

Surplus production has always been present, but it was better managed than the catastrophe we are experiencing at present. Some decades ago, before the general increase in food prices, yams were the staple food of the people of southeastern Nigeria. Forde (1964) records that there was high self-sufficiency in yams, and that surpluses were normally small and disposed of by gifts and sales among kinsfolk. The main redistributive mechanism was a series of seasonal festivities timed to coincide with the harvest period; this closely followed the celebration of the new yams which marked the beginning of the harvest. Such festivities were effective in redistributing surplus production around the group because visitors often returned home with gifts of food such as yams.

One strategy used by the Biase village of Urugbam to control waste is a rule that visitors to their yam festival cannot leave the village on the day of the celebration. And in Afono Erei village the yam festival in-

3. Chambers (1986) disputes this figure at the village level because of studies that relate high postseason losses to marketed grains rather than village-level storage. However, village-level losses result not only from storage problems but also from overconsumption and stress sales.

TABLE 7 Estimated Postharvest Losses in Nigeria 1971–72

Commodity	Postharvest Loss (per thousand tonnes)	Economic Loss (Mil Naira)	Waste (%)
Cassava	1,258.100	32.578	15
Cocoyams	321.750	13.016	25
Yams	1848.249	183.082	15
Maize	104.670	17.438	10
Plantains	50.643	5.046	5

Source: E. O. Idusogie, et al., 1973, "Implications of Agricultural Wastes on Nigerian Nutrition and Economy," *Bulletin of Rural Economics and Sociology* 9 (2) (1973): 255–80. Cited in Banwo Olufokunbi, Gilroy Coleman, and Rex Ugorji, eds., *Record-keeping and Agro Statistics Data Banks in Nigeria* (1986). Ilorin, Nigeria: Agricultural and Rural Management Training Institute (ARMTI).

cludes two days of feasting for about five hundred visitors to the village. This way, people stay longer to visit friends and consume the excess food usually prepared in anticipation of the influx of visitors.

Surplus production today seems to be stimulated more by the need to generate highly storable cash earnings than to provide insurance against accidental losses or for redistribution. There has been a restructuring of consumption preferences manifested in the shift from *foofoo* to gari; the cassava-processing technology has increased the demand for cassava among the Biase; and urban dwellers consume more food. As will become clear, these influences overwhelmed the techniques for regulating the processing of cassava, resulted in an increase in its seasonality, and made it more difficult to regulate increasing urban demand (table 8).

In the years before the civil war, cassava was consumed more in its pounded form, *foofoo,* throughout the eastern part of Nigeria. In this form it was easy for farmers to prevent the release of surplus cassava into the market because cassava was stored in the soil on the farm for up to two or three years and was used gradually according to demand (table 5). This minimized the chances of loss through overconsumption, spoilage, sale at below cost, or consumption by domestic animals. And it takes an average of three to four days to process cassava tubers into foofoo. Women dig up the tubers from their farms, soak them in shallow ponds or huge pots for four or five days to soften them, then strip off the skin and cook the pulp for about thirty minutes, pound it, and serve it with sauce. Foofoo was a standard meal in many homes in southern Nigeria until about 1967, when war conditions left few facilities for its preparation. Three years later, war returnees who had forgotten foofoo affected the consumption patterns in Biase villages.

TABLE 8 Nigeria's Aggregate Food Demand and Projected Supply (per Thousand)

	Demand		
Commodity	1975	1980	1985
Cassava	9,609.124	11,462.165	13,934.414
Cocoyams	953.197	1,136.942	1,382.185
Yams	9,279.809	11,069.264	13,456.842
Maize	1,117.029	1,365.235	1,723.697
Plantains	1,597.643	1,905.692	2,316.754

Source: Federal Ministry of Agriculture and Natural Resources (1979): Food Production Task Committee. Cited in Emman L. Shiawoya, 1986, "Small-scale Farmers, Local Governments, and Traditional Rulers in Agricultural Production." Pp. 45–66. in Adefolu Akinbode, Bryan Stoten, and Rex Ugorji, ed. *The Role of Traditional Rulers and Local Governments in Nigerian Agriculture.* Ilorin, Nigeria: Agricultural and Rural Management Training Institute (ARMTI).

Shortly after the Nigerian civil war, in 1970, there was an increase in the demand for gari, which consequently put cassava into a seasonal "boom and bust" cycle and shortened the time it was stored. The immediate effect was a fundamental change in cassava consumption which tilted preference toward gari, which was more popular with city dwellers because of its portability and the ease of preparing it. Processing cassava for gari may take only a day or two because cassava-grating machines, now available for a fee, have made the traditional method of grating cassava by hand less common. As in similar circumstances where light technology is introduced into traditional production techniques (Byres and Crow 1988), the labor-saving cassava-grating machines encouraged a marketing preference for gari over foofoo and accelerated depletion of the cassava stock. One farmer explained the new consumption pattern: "Before the war I would take two or three balls of foofoo to the market and sell them to get money for salt or crayfish. Everyone ate foofoo. All of us now sell gari because everyone eats gari now; you don't see foofoo in the market anymore."

The lesser time required to process cassava for gari and the longer time for which gari could be preserved (6 to 12 months) over foofoo (24 hours) both combined with other factors, including portability, to increase cassava processing and stock depletion. However, the 1990 fee of three-to-five naira for grating a fifty-kilogram weight of cassava in some Biase villages and the constant malfunctioning of the four machines in Abini and the two in Emomoro/Egbisim ensure that the old system will not be phased out completely.

Because of the cassava-grating technology, gari replaced the more

TABLE 8
(continued)

Supply		
1975	1980	1985
8,631.404	9,745.276	11,025.886
918.605	1,039.316	1,175.892
6,876.821	6,808.327	6,740.515
946.139	1,065.257	1,199.198
1,432.073	1,619.196	2,028.166

popular commodity in rural markets, which until the early 1970s was foofoo; the increased use of gari also meant an increase in the amount of cassava taken out of storage to meet increased consumption and cash needs. Some southeastern Nigerian groups are still unable to produce enough of this crop although the civil war is long over. Back in 1985, the demand for cassava exceeded the supply by about 25 percent; the difference was about 11 percent in 1975 (table 8); this trend is unlikely to change in the future. My older informants reported visiting their cassava farms about once a week before the war years as opposed to an average of two times a week while I was in the field.

Gari is often prepared outside the city and arrives there ready to eat; all one needs to do is pour gari into hot water to eat it with sauce or pour it into cold water and eat it in spoon scoops. Because of the time and effort it saves, city dwellers find it convenient to fit gari into their daily menu. Some migrants returning to the farm who have cultivated a taste for gari have difficulty readjusting to foofoo; this tendency has spread among rural dwellers.

A second factor had a more profound effect in lowering foofoo consumption. For about three years, the Nigerian civil war did not leave rural people the time and place needed for processing cassava for foofoo. All over Biase, people moved deep into the jungle in search of safe places to hide from the war. Whenever conditions were safe for them to sneak back to their farms, they would dig up cassava and prepare gari rather than foofoo. Preparing foofoo entails pounding the product before use, a practice that was unsafe since this would reveal where people were hiding; besides it was unusual to see anyone carrying the cumbersome pounding tools (the mortar and pestle) under the difficult conditions of the war. It was easier for people to grate cassava and transport their gari under the nomadic conditions of wartime southeastern Nigeria. This situation contributed to more Agwagune people consuming gari rather than foofoo, consequently putting rural demands for gari into direct competition with urban demands.

This new consumption habit would not in itself have increased seasonality since rural people could still regulate cassava processing. But urban dwellers suddenly saw relief from their food shortage and regularly made quick buying trips to roadside rural markets such as Abini, Adim, Ehom, and Akpet. This great demand and the possibility for quick profits encouraged farmers to accelerate the rate at which they harvested stored cassava much more than production. The roadside markets became exchange centers receiving bags of gari from the more isolated inland villages of Idomi, Agwagune, Biakpan, and Etono.[4]

Seasonality and Consumption

The cost of food rises by about 20 percent between December and March because farmers clear bush rather than process food; farmers are mainly involved in clearing bush, tilling the soil, and planting. The cost of available processed food, such as gari, rises with diminishing quantity. Although this period is classified as the hunger season because of shortages of food (Gill 1991), coastal Biase communities are less affected by food shortages because they have alternative food sources such as vegetables from floodplain gardens, fish from ponds, and gathering of foods such as bush-mango nuts.

After about the end of July when every crop is planted and there is relatively less farmwork, more time is put into processing cassava for gari. The volume of gari suddenly increases, lowering the price from about one Naira a cup to three cups per Naira in 1990 (five Naira per cup in 1994). The prices of other items, such as rice, fluctuated between one

4. It has been suggested that such increased commoditization may help rural entrepreneurs transcend their scale of consumption limits and serve as an impetus for improved economic circumstances (Johnson and Crow 1988: 144). This possibility is enhanced if rural people are able to regulate food output and processing for a more effective response to product availability and price fluctuations. Very often mechanisms of saving or storage in the form of cash, livestock, or stored crops are necessary if rural people are to experience such benefits (Shipton 1990). Gari prices, for instance, continuously fluctuate according to the volume available for consumption and sale, which in turn depends on what activities the farmer is involved in the farming calendar. But almost always these means are beyond the coping strategies of the farmer.

Naira thirty kobo and one Naira eighty kobo per cup between March and October that year. After about the middle of November, the price of rice drops by as much as fifty kobo in response to the November rice harvest. The price of yams also drops by as much as 50 percent per 7-pound tuber during the harvest months of September to December.

Although the demand for seasonal crops is high, their prices are moderately low at harvest time because of the large quantity available. Fresh maize, for instance, starts becoming available after about the middle of May and is piled up at roadside villages and markets for purchases at low prices. Within the first few weeks of the maize harvest, between May and mid July, an enormous quantity is bought and consumed by both urban and rural dwellers. Enterprising urban traders prudently buy up large quantities which they store and let dry for later sale. Maize loses its moisture content in a few weeks, so very little fresh maize remains in circulation after the middle of July. Dry maize, which is processed into a variety of meals, is then released into the markets. It is more expensive than fresh maize because of the variety of uses which it serves, the possibility of preserving it for a longer period, and the small quantity available. A farmer in Abini sold 80 percent of the maize he harvested in June because he said it would be of no use to him if it dried up; he stored only a small quantity over the cooking area for about four to five months; he would replant it in the next season.

Since keeping dry maize over the cooking area is a way to hold over some maize for later or continuous use, it might seem that our farmers could make a good income by storing fresh maize and selling it in its dry form later. But a number of factors limit this possibility: (1) While every Abini farmer stores some maize, only a few do so with the intention to sell it later (4 out of 30). Most save it for consumption or replanting. (2) The cooking place over which maize is stored is needed between February to October for drying and storing fish, bush-mango, peppers, and various dried fruits and nuts. (3) The demand for dry maize is high in the cities, where it fills a high demand for porridge, but much lower in the villages, where every farmer keeps some for replanting. (4) Keeping maize in storage to fill the high urban demand is hampered by inadequate facilities for storage and transportation.

Lack of storage facilities also affects the enormous quantity of fish available to coastal rural fishermen during the fishing season. Much more fish is available in the rainy season months of May to October, and at this time fish prices are lower. People living in coastal fishing communities consume more fish at this time (an average of 10 pounds per week for our Agwagune sample household) than at any other time (between 5 and 6 pounds during the dry season). Fish is usually dried and stored for sale or domestic consumption, but its availability and circulation dwindle after the rains; it is barely available for sale in some Biase villages between October and January. As with other food items, availability of fish is seasonal and the quantity of fish available for consumption at any time depends on the number of fishponds left to be fished and what storage facility is used.

Smoking is the primary method the Biase use to preserve their fish. Although when fish is smoked, less is consumed immediately, and fish is available for a few weeks to months, fish rarely keeps for as long as two weeks before going bad unless it is constantly stored over the cooking area. An Abini itinerant trader who always sold fish to roadside villagers sent back fish she could not sell to be put back over the cooking area. The fish took long to arrive because the courier had driven one hundred kilometers past Abini to Ikom on some other business. When the fish finally arrived, some had collected maggots and a few more became moldy. Her total loss came to about eighty Naira, but she cleaned up the bad fish and used them for sauce. Holding fish over the cooking area preserves them much longer. In the cities, raw and dry fish are stored for long periods of time in freezers or company-managed cold food-storage facilities, and other methods are available for preserving other perishable commodities. Since no such facilities are available for farmers and because the cooking area holds only a limited quantity of food, we consume more food than we need in the months of plenty so it will not spoil.

Such needless consumption also results from difficulty with moving food from one place to another for sale. More food seems to be consumed at the point of production than is available for external markets, particularly during the months of greater food availability. There appears to be less scarcity of food in the rural communities than in the cities and I noticed no one in the villages who went without food for a whole day. This spatial and temporal asymmetry in the availability of food results from the difficulty of transporting surplus food from rural villages to the cities where demand is greater, but also from the noninvolvement of our men in market exchange (see chapter 5).

Biase farmers who attempt to transport perishable food are hampered by their inability to move the product quickly in large quantities. The average six kilometers Biase farmers must walk to neighboring communities is a serious constraint on transporting goods. Portability is therefore an important consideration affecting the decision to store food for later sale. Maize, for instance, is bulky to transport because it is often stored on the cob (although it can also be shelled before storage) to be roasted and eaten later. Six Agwagune farmers lost 1,200 Naira worth of obiara after waiting five days for transportation to convey their goods twelve kilometers to the neighboring Akpet market. Although this happened in the dry-season month of October when some parts of the trails are accessible by bus transport, the obiara just rotted in the bags because no buses came. The condition of the nine-kilometer trail that links Emomoro/Egbisim with Adim deteriorates during the rainy season and is barely usable in the dry season. There was a weekly average of two buses making the journey from Adim junction, although an average of about sixty people a week used this road to get to the junction. The most noticeable result is that farmers are unable to sell their products outside the community.

Since transportation of food items from the farms to the markets is limited by what the farmer can comfortably carry on his head, the problem of transporting food is a major constraint on economic growth in Biase. Constructing highways through rural areas may help to stimulate the rural economy (Iniama 1983); however, this is a poor way to improve the rural economy if rural people still have to walk between three and ten kilometers carrying fifty-pound loads on their heads. Not surprisingly, a spot check of Biase farmers living by the economically strategic Calabar-Ikom highway does not show them confirming such assumptions. The reasons for this will become clear in the next chapter.

Conclusion

Indigenous technological knowledge is important in the survival strategy of the Biase and has been critical in their ability to cope with unceasing environmental challenges. But the intensification of environmental problems within the past half century has overcome indigenous knowledge and traditional resources, such as land, have been used in less functional ways. While multiple plot ownership remains the most important means of spreading risk, the conversion of arable land to residential land and the loss of labor to migration make this more difficult. Furthermore, although the Biase show a willingness to adopt other systems that will augment their knowledge and supplement food production, they face transportation problems and other problems that make the changes less beneficial. Consequently surpluses from new crops, such as obiara, do little to improve the economy of the Biase because of seasonality, overconsumption where the food is produced, the lack of adequate storage facilities, and the difficulty of getting food to market centers fast enough.

FIVE

The Economy

> Why should people whose mothers are from our patrilineage but who neglect the lineage on the grounds that they were born into another patrilineage have any right to our land? Anyone who refuses to do anything in our lineage because of his obligations elsewhere should also be neglected when it comes to sharing our food. Some people have been fined for one offense or another, but until now they have not settled those fines. The same rule should apply to them and to those who owe financial contributions to our lineage. We can apportion them farmland; but they are not entitled to clear the land for farming until they have settled all their fines and debts; if they farm the land, we'll put them in court.

On this day, Ase Egwa Imo lineage of Emomoro village was allocating farmland to its members. Tempers seemed to be short, because some members of the lineage who were not expected were present. A few of them were said to be less supportive of Ase Egwa in financial matters, either because they had married into other lineages or because they felt a greater link with a parent who was from a different lineage. Many such people had come to get their own farm plots. A young man who felt offended by their presence made sure that the matter was resolved before plot allocation began.

This incident underlines the strategies Biase individuals use in the daily processes of negotiating their livelihood in an environment where patterns of social organization are increasingly influenced by the economic system. While membership in a community or a lineage still appears to guarantee protection against hard times, such protection is contingent upon a demonstrated good standing within the kin group. Many members increasingly have to look out for themselves, particularly where, as will be shown in the next chapter, many members of the kin group do not attend to kinship obligations, help very little in sustaining the domestic or communal economy, or are prevented by cultural factors from maximizing local economic opportunities. In this chapter I explore how these new patterns of social interaction are shaped by some cultural elements intrinsic to the Biase social environment, and the consequences for rural economic growth.

Ownership of Resources

Among the Biase, control of common property rights to land and such components as trees, tree crops, fishponds, and streams is vested at varying levels: in the community of villages, in the patrilineage, and in the individual, with varying rights and privileges. These three levels perform the complementary functions of regulating access and use. But interpretation of these rules differs in relation to membership size, amount of land available, land quality, and resources on the land.

At the communal level Agwagune land is held in trust by the villages of Emomoro, Egbisim, Itu-Agwagune, Okurike, and Ugbem; Abini lands are entrusted to Edodono, Emomoro, and Afifia villages. Each village depends on the support of members of related villages for the maintenance of those territorial rights. Boundaries are marked clearly, with the understanding that land may not be given away or sold without agreement at the level of the community of villages.

The members of the patrilineage ensure that everyone is properly taken care of. Members can lease land to nonmembers as long as such persons fulfill membership requirements such as contributing to the financial needs of the lineage, being present at meetings, or marrying a member of the lineage. Members of the patrilineage jointly exercise custodial rights to their resources, and no one enjoys any special privileges without the consent of all the members. Some patrilineages maintain exclusive control of some resources on their land; in this case resources such as trees and fishponds are not tampered with except by the directive of patrilineage members. If a tree or a fishpond belonging to the patrilineage is located in one's farm plot, then farmwork must be planned so it will not harm the tree.

The power of individuals over land is in nearly all cases limited to their investment in the land and does not include the land itself. Even when a lineage decides to lease land to a member of another lineage for house construction, the land is still the property of the lineage and cannot be leased by the lessee to another person. If the property is vacated in the event of death or migration, the land reverts to the patrilineage, which decides how best to reallocate it.

The same system operates when it comes to farmland. But although a person may not sell his home to another person, a farmer sometimes sells all or part of his farm to another person. Most farms that are sold are cassava farms that are sold to people who process cassava into gari. After the crop has been harvested, the land reverts to the previous owner who will give it back to the lineage, or in some lineages, will farm it again after it is fallowed when the lineage decides to return to that area. Individuals also own the right to invite anyone to intercrop in their farms under mutually acceptable conditions. Despite specific rules that define the terms of the use of resources, regulate the conduct of members, and prevent encroachment by outsiders, problems of interpretation frequently arise. Sometimes the rules are not clear to community members, as was the case during the allocation of farmland by one of the Emomoro lineages.

ODUM: Previously when we have come to break the earth, some people have run to the sections they farmed in previous years and refuse anyone else standing at that section. If we are going to do that today, this whole exercise is going to break down.

TWO ADULTS SIMULTANEOUSLY: No, we cannot let that happen.

EGONG: Listen everyone. These things you're all saying never used to be so. Anyone who takes the amount of land he wants and marks its bounds gets it. Anyone who has no land later is given a place to farm. From Onun Isamo and my father Obin who were big farmers, all had the amount of land they wanted, and no one was ever left without land.

LINEAGE ONUN (ONUN EFIONG): If Onun Odum Aqua is to get his land first, I can be next; or he can get his land after me, it doesn't matter. Then members of the next age set may follow us and we can continue on like that until it is completed. Otherwise, this thing is going to result in confusion.

OBAZI EDET: Onun, if we do it that way then all of us young people are going to end up in the same area. It is not wise to lump all the young men together because if any problem develops between them, it is likely to get out of hand. It may be wise to alternate older men with young men.

What is appropriate traditional practice is therefore sometimes based on the interpretations of individuals who can no longer recall earlier practices accurately. Such confusion may arise from the socioeconomic interaction between members of different lineages with dissimilar practices which influence the behavior of others in the same village. It was agreed that people should stand randomly and take the amount of land they needed to farm. This went on until every plot was occupied. After this five people were reported to be absent, so the lineage elders went back the next day to readjust the plots.

While access and use rights vary little across lineages, the actualization of the stated procedures is usually conditioned by variable man to-land-ratios in each lineage. For example, Igbadara patrilineage of Egbisim operates a system entirely different from that of Ase Egwa Imo. When Igbadara wants to give land to its members, people do not line up and take as much land as they want. Being one of the largest lineages in Agwagune, it has to distribute its available land to the number of people needing farms. The lineage Onun apportions land to each member based on his capability (if he can supply enough labor and money to work his farm) and on the overall need of the individual farmer and his family. Otherwise, land is carefully apportioned to make sure that everybody has a parcel.

Although Ase Egwa has a policy of forbidding people from returning to previous plots, this is not strictly enforced, so, as with Igbadara, permission is often granted for farmers to return to favored plots. Sometimes people inherit plots that were farmed by their fathers. However, if they are unable to plant as much as their fathers did, the lineage Onun may take part of the land from them and give it to people who need more land. Most patrilineages in Agwagune use the Igbadara system of land distribution because only two of the thirteen Odumugom patrilineages own more land than can support their members. As a rule, elderly people who are unable to participate in the allocation of farmland request a lineage member to stand in for them. People are usually careful not to take too much land. On three occasions during "ground breaking,"[1] a few people stated definitely that they did not want too much land and that lineage members getting land for them should get only a small portion.

Land in Abini is obtained through the matrilineage. Sons do not get farmland from their patrilineage but must go to their matrilineage to request land. If they farm in their patrilineage land it would only be at the discretion of the lineage men, who will request termination of use soon after harvest. Women related to the matrilineage only by affinal ties are not given farmland by their own right but are expected to share the land of their husbands. Married and unmarried daughters are con-

1. "Breaking the earth" is a term used by the Agwagune to refer to land apportionment. This term is also reported for the Kanga of Mafia Island, Togo: "'break' a bush field" (See Pat Caplan, "Perceptions of Gender Stratification," *Africa* 59 (2): 196–208).

sidered on the same basis as men of the matrilineage when it comes to land allocation.

Abini and Agwagune emigrants usually retain access to cultivable lands if they wish to. Some return at about the time of land allocation and join their lineages in claiming their own parcels of land. Retaining access to farmland is important for emigrants because of the low wages they receive for their labor, which encourages them to alternate wage labor with farmwork (Barber 1960). Those working not too far from home return every weekend to work on their farms and attend to other matters in the village. Before returning to the city, they take substantial amounts of food back with them to supplement what they can buy with their low income.

Organizing Work

The availability of labor affects the inclination of some farmers to invest in farmwork. During plot allocation for two of the six patrilineages in Emomoro, some people said they did not want to take more land than they could farm because of the shortage of labor. One of the farmers, Ogbada, was a forty-two-year-old man whose crops did better than those of most farmers in the two previous years. Although additional labor was sometimes supplied by his twenty- and thirteen-year-old sons, he said he just needed some rest because he had worked much during the previous season. It turned out that the help Ogbada could expect that year from his sons would be sporadic because one of them was only available during school vacation and the other no longer lived with him. The other farmer was a fifty-two-year-old female household head supporting two grandchildren with little ability to hire additional labor to complete her farm work.

But family labor is not the only form of labor available. Gill (1991) makes a good argument against the logic of blaming labor scarcity when farmers fail to adopt yield-increasing strategies, since it is always possible to increase the labor supply by paying higher wages or hiring migrant labor. The Abini and Agwagune often use migrant workers and local work-exchange groups for expanding the labor supply, particularly in the peak work seasons. Biase farmers cite the high wages demanded by hired laborers as the reason they choose not to hire them. Most farmers do not invest much money in their undertakings; paying hired labor would be a substantial and unwelcome investment. The added difficulty of transporting hired labor between farms and through flooded routes combines with the general inability to afford the going wage rate to discourage the hiring of additional labor.

The flooding of farm routes is a serious constraint to utilizing labor. Weeding periods fall within the rainy season when farm routes are flooded, canoes offer the only means of access to farms, and labor is in greater demand for the slow and difficult task of weeding the farms. Traditionally, women have been maneuvered into being responsible for this task which is a major contribution of women to farming; women also borrow canoes to transport their hired labor or work groups to farms. Ironically, although coastal Biase communities are greatly dependent on water transportation, very few people own canoes (table 6). In Emomoro village only three out of forty-seven women reported owning a canoe, although a few used those of their husbands. Men rarely let their wives use their canoes, because they need to use them for their fishing. People attribute the low number of canoe owners to the high cost of a canoe (300–400 Naira in 1990). Lack of canoes is a major problem for women farmers; they scour the neighborhood every morning and evening looking for canoe owners to help convey them to the farm. At other times, they just wait at the beach and plead for seats in available canoes. The ensuing scramble to find seats regularly results in the capsizing of canoes and the loss of farming tools; but such accidents are considered minor because usually no life is lost. Even when canoe owners charge a small fee (50 kobo) to transport women to their farms, a few remain stranded and spend the rest of the day at home while the work piles up. And sometimes it rains so hard that women have to stay home for more than three days.

Generally it is difficult to do farm work without using additional labor from friends, relatives, work-exchange groups, or hired labor. Our farmers commonly do their own farmwork or join reciprocal work groups to avoid the cost of hiring labor, although social work groups are said to entail more inconvenience than hiring labor (Chibnik and de Jong 1989). Whichever source one gets labor from, problems of flooding and access to the work area are critical considerations, but there is much less worry if the labor is from sources other than hired labor. Relatives and friends understand the economic situation and the environmental conditions and are less likely than migrant laborers to be fussy about bad farm routes. When hired labor is used, it is restricted to one or two workers, to minimize cost. Women from the neighboring Afikpo constitute the dominant hired labor pool during the rains. Rather than negotiate an hourly wage rate, which many Biase women farmers can afford to pay, Afikpo women often prefer to be hired to weed an entire farm plot for a lump sum. The amount is usually much higher than farmers can afford. In addition to paying wages to their hired labor, Biase farmers also spend time and money preparing meals for the workers. Such courtesies tend to blur what would otherwise be a strictly business relationship, giving it the characteristics of acquaintance-based reciprocal groups along with the attendant disadvantages: disagreements over unmet obligations, careless work, or showing up late for work. Although the work quality of hired labor is said to be better than that of reciprocal-exchange or festive groups, there are many quarrels over sloppy work done by hired labor. One involved a case in which a job expected to be completed in a day went on for four days. Although the negotiated wage did not change, the employer was angry that she had to cook for her laborers for three days instead of just the one day contracted for. Such additional obligations are reported to be absent in some other rural economies, because the convenience of not having to take care of the food and entertainment needs of hired labor is its primary advantage over reciprocal work groups (Chibnik and de Jong 1989).

Very few Abini and Agwagune women sell their labor during the peak work period because most own at least four plots which they tend themselves and cannot afford to hire labor for more than one or two plots. In the dry season, some young Biase men and women combine work on their own farms and those of some friends with selling their labor, because of the additional income it brings. In Abini the cost of clearing a farm went from five–six Naira a day in 1989 to eight–ten Naira a day in 1990. The high cost of hiring labor was discussed in a village meeting of Emomoro in January, shortly before the beginning of the 1990 farming. Our women pressured the village to control the cost of hiring labor for farm clearing by setting a limit of five naira a day. But this attempt was unsuccessful because our men used the opportunity to negotiate for a lower cost of gari: they tied the labor agreement to lowering the price of gari, mostly marketed by women, from two cups per Naira to three cups per Naira. The women were dissatisfied because they felt the gari was mentioned just to offend them and to undercut their meager profit. In the end, no agreement was reached; the cost of farm clearing went up to seven Naira, and gari sold at one cup per Naira.

In Nigeria rural outmigration tended to increase soon after the civil war, as war returnees and those dislocated from agriculture sought to fill government job openings (Swindell and Mamman 1990). This situation was intensified by the end of 1987, when Biase found itself within a newly created state structure that was in a hurry to fill up low-level positions in government. About 30 percent of the Abini core farmwork parties (aged sixteen to twenty-five) migrated to schools in the cities or sought wage employment between 1970 and 1988. The effect of this was the loss of the agricultural skill of a large pool of rural labor, particularly in the first few months of migration. Some who were unsuccessful in finding jobs returned to the village after a few months or years, but they exhibited a significant loss of expertise in their performance of farmwork. Explaining this phenomenon, Blaikie (1989) notes that availability of nonfarm wage employment alters the quality of the indigenous knowledge necessary to sustain the community. Wage earners become more interested in budgeting for the next paycheck than in agriculture and learn only what they believe is important to them. This trend has been reported to explain the collapse of agriculture in another part of the world. Younger generations of Spanish Basque farmers abandoned their profitable family farms to work in factories at a lower standard of living; they preferred exchanging a loss of family income for the appearance of dignity in wage labor or the apparent upward mobility of life in the city (Greenwood 1976). Some Biase youths who had briefly sojourned in the city acquired a new status that looked more toward wage labor than farmwork. Many who had gone to city high schools saw rural labor as antithetical to school work, and the title "student" became not only an enviable appellation but a tag of city consciousness.

But the loss of indigenous agricultural knowledge does not take emigrants out of farming; as will be seen later, it is part of their survival strategy in the city to maintain a farm back in the village. Partly because of the venturing into the wage economy and a combination of absence from the village and loss of rural skills, the suggestion has been made that hired labor will replace reciprocal work groups (Erasmus 1955; Moore 1975, cited in Chibnik and de Jong 1989), because migrant farmers will be able to pay for labor rather than exchange their own. Understandably there is little support for that opinion, because the economic circumstances of rural peoples rarely improve as a result of their involvement in wage labor. When they do improve it is more often the case that the improved economic conditions also bring about the ability to command the labor of both kin and nonkin (Webster 1990). Some of us who hold jobs that pay more than the minimum wage are able to get even close friends and age mates to help us with our farmwork. In some cases we send instructions from the city specifying what we need done.

Such labor is done with expectation of drinks, food, or conveniences from the city later when the owner of the farm visits the village, but there is rarely a direct cash payment for the labor. This affectionate transaction between persons involved in exchange of services characterizes the economic system of the Biase and is significant in shaping our entry into, and interaction with, the money market, which is now our dominant system of exchange.

It seems, however, that those who have emigrated to work elsewhere contribute very little to the Biase economy. Only 4 of the 269 households in Odumugom and 16 of the 437 in Abini reported that family members in the city sent them money at least once a month. Stories were told of some Biase emigrants who rarely visit their villages because they want to stem the financial demands of family members. Many reported that family members or friends visiting from the city often asked for money to pay for transportation back to the city. The low-wage jobs in which most Biase find themselves may partly account for this. Some emigrants report that the demands of city living overwhelm their obligations to family members back home.

Distribution

In Biase we attribute our underdeveloped economy to the lack of a market to stimulate the production and distribution of goods and services and the lack of an all-season road to arouse the interest of prospective visitors. It is significant that we identify elements classified by some social scientists as likely to result in rural misery as holding the key to our development. This perspective is embedded in a historical precedent, when trade and river transportation brought the Biase economic power and fame. Until about 1970, river transportation was not only the most feasible form of movement in southeastern Nigeria but also the politically and economically most rewarding. All the important actors in the Cross River palm oil trade—Agwagune, Efik, Umon, Afikpo Igbo, and the British—had a coastal operational base, because the "great link between Calabar and the upper waters of the Cross River was the water highway" (Nair 1972: 245). The Biase were principal actors in the commercial ventures that operated along the Cross River coastal stretch (Attoe 1990; Nair 1972). When the British arrived in the area in the middle of the nineteenth century, they encountered a great deal of difficulty in their plans to control trade along this river route. Nair writes that "towns which were considered useful were in the habit of subjecting others to continual depredations. The important town of Agwagune is an apt illustration of this dilemma. The British were keen to support Agwagune against its rivals since its people were venturesome traders who traveled long distances in their canoes to bring down oil, rubber and ivory. As an agricultural and industrious people the Agwagune would be useful in opening up the [Cross River] country as they had the necessary trade links" (1972: 248).

This dominance gave the Agwagune a flourishing trade in yams, palm oil, fish, and vegetables; the difficulty of arriving at our villages was not our problem but that of other communities who needed our goods, because transactions were limited to and easily undertaken with riverine communities. Noncoastal villages were unimportant in our calculations of economic dominance because the lack of a water transportation route made such communities less-attractive trade partners.

Consolidation in trade was further enhanced by the geographic location of the Biase. Partially surrounded by jungles, streams, and rainy season rivulets, the Agwagune were strategically isolated from the neighboring villages, which made it difficult for outsiders to attack them and limited encroachment on their resources. Ironically, these same geographic conditions are today important in negatively restructuring the economic fortunes of coastal Biase villages.

Increased siltation over the years has raised the level of the Cross River bed, making navigation difficult from about October to March. Vast stretches of infertile sand intermittently break up sections of the river, making it less navigable even for the small canoes we use. Biase beaches at Emomoro, Egbisim, Ikun, Itu-Agwagune, Okurike, Etono, Ugbem, and Umon have experienced landslides that have destroyed once economically important beaches which provided anchor points for large trade canoes. Today, trading thrives only in distant markets far from the coast; highways which skirt the edges of the jungle and leave the Biase at the fringe of the urban experience are the major means of transportation. Except in Calabar and Afikpo, where a few city amenities exist, the former economic giants of the Cross River are today mere shadows of their old selves. And today the Agwagune live in economic and political obscurity.

It is against this background that the Biase say that transportation and a market would "get them off the ground." This local perspective then, will form the basis for examining how participation in a market system and the availability of good roads interact with gender relations to affect rural economic growth.

Anthropologists have long known that the distance to a city and its accessibility or location near a market town have a dramatic impact on traditional agricultural strategies (Uyanga 1980; Barlett 1982; Iniama 1983). Economic theory articulates this position best in stating that the growth of an area is directly determined by its ability to market its products and services outside its boundaries (Uyanga 1980). This model stresses the need for a usable transportation system linking the local population with external markets since the rural market is too small to generate the volume of trade needed to effect growth. In his analysis of the economic effects of Nigeria's rural transportation on development, Uyanga attributes Nigeria's slow economic growth to the lack of good access roads linking the rural areas where most of Nigeria's food is grown to the cities (1980). A later study of the impact of the Calabar-Ikom highway on agricultural development concluded that the road has had a positive effect on agricultural productivity in its service areas, which include the Biase (Iniama 1983). In her study of the Costa Rican community of Paso, Barlett shows that before the construction of a highway, only a few households grew large amounts of coffee because of transportation difficulties. Since the highway was constructed, families are making additional income by transporting sacks of coffee to buying stations near the highway (1982).

The assumption that road transportation will open up rural areas for development is a standard one that is popular with both scholars and rural peoples. But difficult access from the farms and the intervention of certain cultural attitudes which are examined below may make it difficult for a good transportation system on its own to improve the economic circumstance of any community. Although the town of Abini has enjoyed excellent road transport since 1970, some informants described its economy as one of the worst in Biase. On nonmarket days during this study, an average of only nine cars stopped at Abini to pick up travelers, as opposed to an average of twenty on a market day. The presence of the road does not encourage Abini farmers to increase food production because the availability of transportation to market centers is not reinforced by the ease of transporting food from the farms to the home, nor does the highway guarantee the movement of surplus food once it arrives from the farms. If such locale-specific circumstances, including the condition of the farm environment, are not considered, it is possible to speculate that farmers will benefit from the presence of roads in the general area.

Abini also suffers the anomaly of sharing good transportation with older towns. While the Calabar-Ikom road may increase agricultural production around certain towns in the area, such gains may be at the expense of other towns within the zone (Lipton 1985). For instance, Ugep, which straddles the highway, was already experiencing economic growth before the highway was constructed. Villages located about twenty kilometers inside of Ugep would actually be hurt by the highway because they would lose business to the equally accessible bigger food market in Ugep. As a result, the Abini located twelve kilometers from Ugep and historically outside the mainstream exchange network have not increased their entrepreneurship because of the mere presence of a good road. They still need to get their products to centers of demand and to experience interest in trading in the community.

The Abini attribute their lack of development to a general unwillingness of community members to help each other economically. This perspective may at first seem simplistic, but is has some credibility. Abini sees itself as belonging to one large matriclan. The three Abini villages have Inun who are under one Onun of Abini. If there is a project such as building a community center or constructing a farm route, the villages work together; one is not left to carry the burden at the expense of another. This spirit pervades the structural base on which Abini development may occur and has fostered what has been described in a different context as an "economy of affection" (Hyden 1980). In the present context, it describes a condition in which the Abini believe that for any of them to be financially successful, there must be a substantial moral and financial commitment from other community members. Ukpabi was a farmer with a large yam farm who was making little money from his sales. He reported that he had intended to take his yams to Calabar because he was sure they would bring more money there than in Abini. But he abandoned this idea because he did not get other people to join him in the venture, since if he was successful, all those who had refused to join him would come to borrow money from him and would bankrupt him. If cooperation is not forthcoming, there is the assumption that the project would not be beneficial to the village. It is within that perspective that we have to understand why the Abini believe they cannot better themselves without getting help from their neighbors.

No doubt such beliefs, which support the reluctance of some rural people to take economic risks, are also amenable to other explanations, including the "safety first" principle of the moral economy approach. That is, the vagaries of subsistence conditions make poor farmers "refrain from initiating changes in local systems except in desperate response to the destruction of traditional social safeguards" (Parker 1988: 181). So although Ukpabi believes he would make more money by taking his produce to Calabar, he is not prepared to face a possible social conflict resulting from his refusal to share his fortune with importuning relatives. Preserving social safeguards is at the core of Biase economic arrangements but it does not in any absolute sense determine individual economic choices. In this case, Ukpabi, like other individuals, could leave the community and operate his business elsewhere without recriminations from relatives. On the other hand, he is also making a rational economic decision by weighing his profits against the time he would lose from farmwork if he traveled to Calabar. Besides, there is no way to guarantee that he would make the expected profit or even be able to sell his yams if he traveled to Calabar. Of central importance is the fact that rural people cherish prosperity but not as much as they value social morality, because while individuals have self-interest in economic prosperity, the economic decisions they make are not isolated from traditional values. We must understand the myriad of noneconomic factors that are pivotal in the economic decisions we have to make in order to understand the interplay of our traditional values and the desired economic goals of individuals, specifically, our involvement in the market economy.

The Market

Indeed, the most significant feature of the production of farmers living within a state polity is their involvement in the market economy. Like most rural groups, the Biase are now fully monetized but are unable to cope with the risks and uncertainties within the market structure on which their livelihood depends. I noted earlier that profit making has become a major consideration in production decisions, yet Biase farmers expressed more willingness to sell at low prices to itinerant traders than to walk long distances to neighboring markets. Willingness to sell below production cost does not imply ignorance of prevailing prices but is a conscious decision borne of the knowledge that although outside market prices are higher, the trader is unable to convey enough to the market to cover her overhead. Besides, since Biase villages produce similar foods, there is often a glut of items in rural markets—which is why they are so attractive to city dwellers—and it is often difficult and wasteful for traders to carry home items they cannot sell.

The nearest market that Emomoro/Egbisim women attend frequently is the Abanwan market. This market is convenient not only for its relative proximity (5 kilometers) but also because Afikpo and Arochukwu traders bring in various nonfood commodities which are lacking in many Biase markets. Because neighboring communities cultivate identical items, our rural markets are flooded with staple food items. Since the bulk of buyers are also sellers, much food is wasted because there is little demand for it in rural areas. Gari, for instance, is a staple food of many Nigerians and is processed by almost every woman in Biase villages, as well as in other rural communities in southern Nigeria. Although there is a steady demand for gari in the cities, there are few buyers of gari in rural markets because almost everyone is a producer. But urban demand does not seem to have any significant effect on the price of gari or other long-storage food items. The abundance of gari, for instance, did not lower its February price of about 1 cup per Naira. This is partly because the possibility of storing gari longer than some other food crops gives farmers some flexibility in holding onto desired market prices. However, such flexibility is never sustained for long; prices fall quickly for, as I noted in the preceding chapter, rural storage and preservation conditions do not give much latitude for choice.

To prevent food glut, some communities operate a zonal trading regulation that specifies who will carry what goods to the nearby market. Abanwan community and Odumugom of Agwagune classify their villages into two participatory zones. On some market days, the women of one of the zones are the only people permitted to take regulated items (gari, obiara, yams) to the market; women from the other zone may still go to the market and sell with nonregulated food items. During the first two months of my research, I did not notice or hear report of anyone who breached the zonal trading practice. But later investigation revealed that zonal trading was not being strictly followed, since there was a significant number of persons from both sides taking regulated items to the market on days they were not permitted to. Some women admitted that the rule has always been breached although no one has ever considered a violation serious enough to warrant a reprimand. The existence of the rule, however, helps to minimize food glut and to stabilize prices to some extent.

Zonal trading rules do not apply equally to all traders. Abanwan market is a high-volume area for goods and traders especially because of the participation of Igbo traders, who are desirable partners because they sell most of the nonfood commodities. As a result of their preferential status, Igbo women who trade in perishable commodities are unaffected by zonal trading rules. This undermines the intended effect of giving Biase traders the maximum opportunity to exploit the market.

Such inequitably applied rules point to one of the ways rural people try to cope with processes they are unable to master, resulting in further weakening of the economy.

Although Polanyi et al. (1957) have expressed dissatisfaction with using conventional economics to explain the operation of premarket economies, the increased monetization of poor communities continues to expose them to market relations and attempts to maximize profits help to shape their behavior; consequently, we can no longer leave the economic behaviors of poor communities strictly to substantivist analysis. My concern with market relations is neither an endorsement of the perspectives of mainstream economics nor a rejection of it. Rather, it is based on the assumption that inquiry into rural economic growth will be more rewarding by combining both formalist and substantivist approaches.

Associating imperfectly with the market economy may result in dysfunctions that encourage the opinion that rural people should be isolated from the market economy. However, Collier et al. (1986) have noted that rural economic dysfunctions do not appear to be a consequence of unfavorable or exploitative integration into market processes, but instead stem from lack of integration into such systems. Another writer has suggested that access to productive resources is important for sustaining livelihood (Berry 1989). This access depends "on participation in a variety of social institutions, as well as on material wealth and market transactions" (1989: 41). In other words, the long-term effect of the current peripheral relations with the market is to make the farmer/trader a more sophisticated participant. According to my research, there are strong cultural elements intrinsic to Biase social structure that appear to offer little hope for such optimism.

On the market days of neighboring villages, the Biase traditionally suspend other occupations in order to participate in market activities. But their level of participation today is still guided by the desire to nurture traditional friendships in a monetary economy that characteristically operates without love. Those markets some Biase villagers traditionally visited have experienced economic changes that tend to place them in a higher market hierarchy than many Biase villagers are used to. For instance, the markets in Ediba and Ugep are now located on major highways running from the south to the north of Nigeria and attract the more sophisticated Igbo and Ibibio traders who compete with Biase traders. This tends to increase poverty because poor farmers lack the knowledge of market mechanics, since they are largely illiterate, lack experience, and lack technical skills for negotiating at more than a "subsistence" market level. When I asked my Agwagune households to estimate their profit over eight consecutive visits to the Abini market, two traders admitted making a profit of between 2 and 5 percent, nineteen could not tell whether they made a profit or not, and some of them expressed relief that they were able to sell their wares at all: "better than throwing it in the trash." No one admitted that profit was any motivation for going to the market; many people indicated they just wanted cash to purchase items they do not produce locally. This imperfect relationship is supported by the market structure of the Biase and the cultural ideals that direct its management.

Periodic Marketing

Structurally, rural markets are characterized by their temporal and spatial periodicity, such that "market meetings are separated by marketless days" (Smith 1978: 12), and also by distance; this market structure is familiar to visitors and scholars working in Asia, Latin America, and Africa. Hill (1963) has identified seven types of periodic markets in West Africa: some have a 2–8-day periodic schedule; a 14-day schedule has been reported for Ghana; and most East African and Latin American markets follow a 7-day regime. Here we will use the example of the Biase who operate a 4-day regime to examine the cultural dynamics of the factors related to Biase participation in market activities. The focus will be on understanding the economic consequences of use of alternative market days and differential gender participation in marketing.

Most people stay away from farms on Biase market days so they can attend the market. But rather than participate in marketing, many of our villagers engage in activities that keep them away from the market. On nonmarket days almost everyone is in the bush; during the peak season most homes are shut between eight in the morning and about five in the evening. But on market days, Agwagune villages prohibit people from going to the bush until the market is over. Our market days are therefore convenient for fulfilling communal obligations neglected because farmers spend most of their time in the bush: men assemble for the town meeting, women hold their weekly meetings, and younger age sets are sent out to work on some community project. The argument in favor of this arrangement is that the market never really begins until about noon, when those assignments may already be completed. If that were so, it would translate into good communal management. Women I spoke with, however, said at the end of market day assignments they are usually too tired to go the market. Yet they make an effort to visit the market for a short time in order to avoid paying a fine. Traders visiting from neighboring communities complain that the Biase make it difficult for them to sell their goods. Afikpo traders who bring in essential nonfarm items arrive at Agwagune markets early and wait till late in the afternoon before Agwagune women begin trickling into the market. Market activities go on for about two hours and then the women disperse to visit their farms and get some food. The only traders who stay after 2 P.M. are those visiting from outside. One visiting trader said he stays longer to make up time lost because of the market starting late. He does not usually sell anything when he stays late but he justifies the time and money he puts into traveling to Agwagune by buying some food items that are less expensive there than in Afikpo market. Some disillusioned traders take their goods elsewhere. Odumugom informants said the number of traders visiting from outside has diminished because there used to be at least eight traders selling clothes and about eight others selling other nonfood commodities. In the year of my study there were never more than five visiting traders. Okurike market, which was a major exchange center for Biase traders until about 1970, has now lost this importance. Visiting traders who made up the bulk of participants have left for more convenient markets situated along the Calabar-Ikom highway, leaving Okurike virtually stagnant in the hands of only its women traders.

Biase markets are small and give the customer very little to choose from. Items are assembled randomly at places the sellers find convenient. Usually women make up the bulk of the sellers and buyers in any of the periodic markets. On three sampled Emomoro/Egbisim market days, there were 139 women and 2 men, 110 women and 3 men, 141 women and 5 men, respectively—an average of 130 women and 3 men per market day. None of the men on any of those days was from Agwagune.

Although informants said that the Agwagune rule requiring people to attend the market also affects men, it seems to have been designed with women in mind. The rule specifies that any woman who is registered in the women's meeting must be in the market on market days whether she has something to sell or not (plate 22), and that men who have yams may also take them to the market. Women are careful not to break this law and always go to the market before going off to the bush to do other chores, because they are regularly fined in their weekly women's meetings if they are absent from the market. Men rarely participate, and there was no instance of a man being fined for a breach. Anthropologists have long noted the greater involvement of women than men in the market economy (Mintz 1971), but where the condition persists as a cultural ideology, the effect on the group's economy can be critical. During one general village meeting the question of the absence of men from the market was raised when a woman shouted in the direction of the men: "Is there any town in this world where only women are traders and the men never join them in trading? I am asking because that is why we are not getting off the ground. No strangers are coming to buy our things because we do not have a strong market supported by both men and women." The reason some men gave was that at the time of my interview (April), yams were no longer available for sale because men were preparing them for replanting. If they had just harvested yams, some of the men would have been at the market selling them. Since they had no yams, it was unnecessary for them to be at the markets. The months following my fieldwork revealed that this was not the case, and that men regularly stayed away from the market even in the yam harvest season of September to January. Asibong, who operates a small store, admits he never takes his goods to the market on market days: "Everyone knows my house and also that my store is inside here. If I go to the market, I am going to sit there and argue with all those women instead of selling my things. Men are not like women; we don't plant much obiara, maize, pepper, or cassava, or process gari. It is the women who do that. The market is for them to sell those things to other women."

Abini men said they did not have to go to the market because their women would usually take their produce to the market for them. If they had yams to sell, the women would take them along with their cassava, obiara, and vegetables. Although Abini market is located right on a busy highway, most of the men traders are visitors bringing nonfood items from faraway villages located along the highway or city dwellers stopping by to buy food at cheap prices.

According to Onun Unoh Aquah, lack of participation by men in marketing activity is a new phenomenon that has both a political and an economic explanation. The political position of our men has been considerably weakened by the loss of the eminent position we enjoyed in the days of the Cross River oil trade. Trade friendships established with noncoastal rulers, which included helping them sell their yams at the coast, are no longer important, since state highways fulfill such needs much more easily. Consequently there is little need for noncoastal communities to maintain political links that are no longer vital for trade. But the Agwagune still see themselves as political superiors in spite of this weakened political status and tend to maintain such myths rather than deal with people traditionally classified as inferiors. An elderly informant reminiscing of the past said, "Look at me. I stayed away from my farm because I was afraid the Adim would kill me. I would never have imagined that in this lifetime someone from Adim would stop paying the Agwagune yam rent for our land they occupy." This search for old glory is important in understanding why Agwagune men tend to be unenthusiastic about participation in the market.

And the yam trade has lost its importance, because of decreasing demand in barely accessible areas. Informants recalled that before the 1967 Nigerian civil war, our men would stack about a thousand yams in large canoes and set off on trips to villages up and down the river. Such trips not only gave us enormous wealth but it also brought us respect from noncoastal communities that made long trips to the coast to buy our yams. At the same time, Biase women's involvement in the economy depended heavily on the decisions of their men. Very often it was limited to helping convey palm oil and yams to the canoes or to neighboring markets. The man decided if and what the woman would buy or sell even at the subsistence level, particularly because men sought to minimize women's contact with the numerous strangers trading on their shores. Although women participate in the economy much more today than in the oil trade years, they are still constrained by cultural factors from effectively exploiting market opportunities.

During the palm oil and yam trade, the Agwagune expanded their clientele by taking their goods to far-off coastal towns. This is no longer the case. Abini informants said they do not visit other markets because they are self-sufficient; besides, people who need their goods can come to their market. Equator's Father, one of my informants, put it this way: "Where would I go? My cassava is there, my wife's vegetables are there. At Christmas I cross the street and buy two head-ties for everybody. My palm wine is there in the forest. It is people like you who keep running to other people's countries because you don't know how to work in the village. You go around with your book and pencil, I carry my machete and my gun."

The demise of the coastal trade and the withdrawal of men has left women to manage an institution whose structures no longer support its terms of operation. That is, a previously male-centered economy has failed to come to terms with its new managers, who are constrained by cultural attitudes and traditional liabilities from expanding its potentials. The change in women's social and economic status since the collapse of the Cross River oil and yam trade has made them not only the main food producers but also the primary entrepreneurs in Biase, as well as the major financiers. In some ways, this has affected the maximization of periodic marketing. Most of the items seen in the markets are subsistence crops because since women are the main food producers and sellers, they often exchange only what they produce. This keeps the Biase economy at the subsistence level with few alternative income sources, because the reliance on subsistence income makes farmers predominantly poor (Collier et al. 1986). Faced with a social organization that demands they take care of husbands, children, and parents, women attempt to manage the town's economic affairs along with domestic affairs and have little opportunity to expand their economic activities.

This is a familiar experience across the globe as domestic responsibilities contest with farm and nonfarm labor demands to lower the economic performance of women. Humla women mountain farmers of northwest Nepal, for example, juggle childbearing and childrearing with the high labor demands of mountain agriculture and, like some of the rest of the world's poor women, spend more time on agricultural work than on child care (Levine 1988: 235). It is also reported that the market women of Antigua, Guatemala, confine their operations largely to the domestic market or day trips to the capital because, as is familiar among the poor, "women's routines of travel are more restricted than men's, and their market careers may be interrupted or delayed by childbearing" (Swetnam 1988: 334).

Few Biase women have the amount of free time for traveling from one periodic market to another which is a requirement for the successful operation of periodic marketing. For such markets to provide any economic benefit to their practitioners, consumer demand must equal the cost of producing and offering the goods and services at a central place. If the consumption is low, then the traders have to be mobile in order to get to more consumers and stay in business (Fagerlund and Smith 1972). As women spend more time in farm and domestic labor, there is less time for participation in the market, and also fewer commodities are brought to the market. Some women go the bush about seven in the morning and return at about noon to stay with their children, using the rest of the day to do domestic chores. This cuts their average farm labor day by over 50 percent because they generally start returning from farms at about 6 P.M. Two mothers of infants we observed lost an average of four daylight hours of work every day they came back early from the farm. After I observed them for ten days, one of them had lost thirty hours and the other thirty-seven daylight hours. This is hardly beneficial for a culture that survives on the work of its women, because the time needed to service the communal economy is used to sustain the domestic economy. The consequences are that women will produce less for subsistence and be left with no surplus for the markets; they will become more dependent on husbands who earn less than wives; there will be a few less women in the markets; since the women's association bears much of the financial burden of the community, its total income will diminish because some women are unable to meet the association's financial demands. The consequences for the community include a loss of developmental impetus from the major financier.

Even with these constraints, women do manage to attend neighboring markets regularly, although the volume of their business is affected by the amount of goods they can carry from home to market destinations. They carry goods on their heads over long distances. For the Agwagune, a major limitation during the dry season is a two-mile walk across hot sand to the pier, where a boat is used for a nine-hundred-yard water crossing; after crossing the water there is another four-mile trek from the other bank of the river to communities west of the Cross River (Biakpan, Etono, Ikun, Erei, and Afikpo). This trip is less cumbersome in the rainy season when traders board canoes at the beach and disembark a few meters from the market location. But the trip is also hazardous. The rapid and swollen river makes traveling at this time risky. My informants reported that there were three cases of capsized canoes during such trips last year. Market goods were lost, but no lives were lost in any of the incidents because every Agwagune person learns to swim when very young.

Agwagune men say they have little interest in marketing because the Biase do not have such state-provided amenities as good roads which their neighbors enjoy. This idea was expressed at a patrilineage center discussion about the better economic position of the Afikpo and some of the Abanwan traders. According to Edet Ebok, one of the Inun in Egbisim,

> People don't come to our market because they cannot transport their goods here easily. In a place like Akpet, vehicles drive right into their markets every market day. But it is not possible for something like that to happen here since we are surrounded by water. In the dry season when the govern-

> ment should repair the road, nobody is there to remind them. Every year they will say somebody is going to do it, but we have not seen it. If you are a trader, will you come to a place like this to sell your things? They (traders) know that we have no money here. We are poor, so they go to Afikpo and Abanwan and Abini to sell their things.

Onun Effime Agbam of Igbadara added,

> When we go to other markets, we can carry very little because there are no cars to take our things to Akpet or Abini or even Ugep. That is not how people trade in the market. If people trade like that then all the Igbo traders will not be big. But when the Igbo man goes to trade, he carries everything in a pickup (truck) and goes to sell it. After the market he will make enough money to send his son to college. Our children don't go to college because we have no money. They don't read books because they are always on the farm. We work on the farm too much but we don't have any money. Last year there was nobody to buy my yams at a good price. If I took them to a big market like Ugep maybe I would make plenty of money. But will I carry two hundred yams on my head to Ugep?

Before the opening of the Calabar-Ikom highway in 1973, the town of Ugep was in the marketing ring of the Biase; Agwagune and Abini farmers went by bush trails to attend Ugep market, which was "the largest village market for some twenty miles around" (Forde 1964: 40). Forde reports that women traders arrived in the market in considerable numbers with dried fish, vegetables, and cocoyams, and some of them remained in the village for several months while traveling out frequently for supplies. This early initial advantage contributed to the enormous economic progress that Ugep has experienced, and as was suggested earlier, it also encouraged the withdrawal of businesses from other Biase markets.

A related reason for Biase economy lagging behind those of their neighbors is that although groups such as Ugep shared the mutual disadvantage of bad roads with the Biase, they were also well situated to correct some of their transportation problems. This factor is referred to here as a *mutual disadvantage correction* and describes the condition in which some villages within a market ring experiencing a transportation disadvantage in common undergo a correction in their conditions. Communities within a marketing ring share a mutual disadvantage of unusable roads, and exchange is restricted to a limited quantity of goods that can be carried on the head over a limited walking distance. But in the dry season, this disadvantage is corrected for some communities if their seasonal roads are more usable than those of their trade partners; this enables them to increase trade with other market centers, while their neighbors are unable to do so. Some traders in the advantaged communities use bicycles as a means of conveying more commodities to markets. As poor road conditions temporarily interrupted mutual exchange between Biase villages and their neighbors, Ugep continued to trade extensively with other groups outside the market ring of the Biase and Afikpo traders concentrated their trade with other Igbo villages that were more easily accessible than Biase villages. And rainy season waterways make it easier for some villages in a marketing ring to use canoes for trading with neighbors outside the ring, while others are isolated by their difficulty in traveling through mud and flooded bush. The advantaged communities develop into middlemen as they buy commodities from the disadvantaged towns and sell them at a profit to other market centers. The high prices outside the Biase market ring encourage some traders to pay daily visits to disadvantaged markets and buy commodities even on nonmarket days and subsequently sell them at a profit in other markets. Therefore an unequal relationship develops within periodic market rings as some communities operate in the market ring only peripherally, while others benefit through a corrective factor that is advantageous to them and less so for others within the ring.

There are strong reasons, of course, for arguing that periodic marketing as practiced by some rural villages is conducive to the economy of those villages. For instance, it restricts the flow of food items out of a community where food production is largely for subsistence. Should our villages operate daily markets, the quantity of food would be likely to drop drastically and might result in inflation. Since we are involved in other subsistence and social activities, such as fishing, house building, hunting, the gathering of fruits, and leisure time, life would be hard for us if those spheres were neglected. Rural subsistence work is a continuous task. Abini women say they get despondent if they have to stay home because of illness. An Agwagune woman who broke her neck on the way from the farm during the rainy season kept expressing the hope that her pain would subside just enough for her to go back to the farm. In these circumstances, it would be difficult for the Biase to set aside enough time to operate a daily market. Yet daily marketing seems an appropriate model for the Biase.

Periodic marketing, with the spatiotemporal relationship that characterizes it, is explained as arising because of a need to prevent intermarket competition. Regularity in spacing allows markets within an exchange ring to serve as the nuclei for neighboring villages, and a temporal schedule permits each market to operate at a maximum without competition from another market within the ring. The popular explanation for this spatiotemporal dynamics fits the assumption that it is better for us to rotate congregation around production points since we do not produce enough to warrant a daily market. Such a situation may have been necessary many decades ago when our social and cultural lives did not extend beyond our familiar environment. Today the effects of exogenous trading are significant since we take our commodities to neighboring markets on a daily basis without waiting for culturally designated time and space zoning. If Biase women cannot wait for four days before trading their goods, that suggests that we no longer find periodic marketing an effective or convenient cultural practice for fulfilling our economic needs.

Daily trading has become a standard practice in rural communities, but the fact that it is done imperfectly underlines the necessity to make rural people more knowledgeable about it. At the end of the day's work, some villagers sell their major subsistence crop to other members of the community and then go around every evening complaining and telling members of the community what they need. They often walk into the homes of people who have those items for sale and buy them. And men return with their fish and sell most of it; after a couple of days their wives complain there is no food because there is no fish in the house to cook. They then repeat the process of looking around and buying fish from the neighbors to whom they sold their fish. Of the farmers we questioned, 53 percent of the women estimated that they sold one-half of their monthly production and that they had to buy the same item (usually gari) at least five times in the course of a month. Men did so more often; 90 percent of them sold about 82 percent of their catch. Forty percent of the households of such men spent 20 percent more than their total earnings from fish to buy fish for the household; their wives said the money rarely came from their husbands. The need for cash is the primary motivation for this exchange, but the point is that because there is a supply-and-demand interaction occurring on a daily basis, the needs of rural communities can no longer be satisfied by the present structuring of periodic markets.

The structure for daily marketing already exists in our villages. Finan (1988) reminds us of noncontractual agreements that exist in rural communities between buyers and sellers that tend to foster long-term market relationships rather than short-term price maximization because the participants perceive indirect benefits from foregoing direct short-term advantages. He says that "to mitigate the risks inherent in a marketing enterprise, the actors interpret impersonal, competitive relationships in terms of familiar social and cultural texts, where trust injects a sense of predictability into outcomes" (1988: 695). So the local storekeeper continues to part with his goods without demanding immediate payment; his customers, as a result, continue to patronize him even though his prices may be unusually high. But the personalization of this relationship fits perfectly into the simple patterns of rural existence, and according to Finan, is part of the social process of everyday living. And the postponement of payment releases funds for other income-generating transactions that raise the overall income and create a surplus that will eventually absorb the debt.

A related component of the market in association with farming is the development of distribution channels. Dannhaeuser (1987) discusses the role of industrial consumer goods as incentives for small farmers to improve output, because the poor are not so isolated that they are unaware of or uninvolved in the distributive network of industrial commodities. He offers an incentive goods argument which predicts unlimited consumer demand by those who produce as providing an incentive for them to increase production if as a result they have prospects of increasing consumption. The rationale is that a man will do more work depending on the additional goods he can get in return for his increased effort (Dannhaeuser 1987). If the desire for consumer goods stimulates agricultural production, this suggests that communities shut off from the possibility of buying things other than their most basic needs will continue to limit their economic goals to familiar transactions.

This is particularly true in the usual cases where 80 percent of people in low-income economies live in the rural areas, are independent as far as their own food production is concerned, and have neighbors who are equally independent. They also have limited markets for their farm surpluses, and existing markets are not well developed and prices are invariably low (MacDonald 1989). The average farmer then sees no value in reassessing or updating his marketing strategies if there are no markets, or no markets with attractive prices. Such limitations are bound to stifle economic growth and intensify poverty. However, there are arguments against this strategy, especially one that sees the market involvement as continuously putting rural people in debt, thus increasing their poverty. My own data will explain this point of view.

In the past ten years, some Agwagune traders have lost their businesses because they could no longer afford to stock their shops. In 1980 Emomoro/Egbisim had four men traders in basic household needs such as soap, cream, towels, and dishes. Ten years later, there are only two left. Within the past five years, the remaining two have increased work on their farms rather than depending on selling nonfarm goods as they did previously. One of them, Asibong, observed that "Agwagune people buy everything on credit. You have to give them credit because if you don't you will not be able to sell your goods. If they need sugar, they will come and take it and say 'I will give you the money when I return from the bush in the evening.' If you refuse, your things will stay there and turn white with dust."

The result of selling on credit is that some customers wait for months to pay after making a purchase, some are unable to pay, and others forget the debt entirely. A onetime petty trader reported that if payment of a small amount, say fifty kobo or two naira, was withheld for too long, she would write off the debt. An amount going up to five naira is considered large, and people will sometimes spend the same amount to take debtors before the local arbitration units. It is not likely that businesses operating within this economy can support their owners; eventually they fail and the debts remain uncollected. Such business failures are mostly experienced by sellers of cigarettes and drinks such as beer and the locally brewed gin, *ofofop*. People just walk into the seller's home and take a cigarette or two or ask for a few bottles of beer or the amount of ofofop they want. Usually they seem to be in a great hurry and promise to return with the money in a few minutes. Some return a day or two later to ask for more drinks and ask for the new account to be added to the old one. Edodi sells beer, ofofop, and kola nuts, which are all popular consumption items. One day I walked into her home in the company of Obazi, one of my key informants.

EDODI: Obazi where is my money? I have been asking you for two months now to give me my money but you have refused. You men in this village do not respect yourselves. Where do you think I will get the money from to keep buying you drinks to owe me for if you don't pay me what you already owe me? Your money has come to seven Naira. After this don't ask me for another drink please.

Obazi: (He smiles, turns to me, and whispers) You see, this lady is the best woman in this town. That is why I always come here to drink. She does not know that I have her money right here in my pocket. Just wait and see how I am going to surprise here. (He addresses Edodi) Mama Edodi, yes, you are correct, and I am going to show you that I don't like owing anybody even one penny. It is this my friend here that I want to buy ofofop for because he is my friend. But if this were my ofofop would I sell to you if you owed me seven Naira? No. If I don't have money, Iyam here will help me because he is a man. Here, take three Naira. My money is now four Naira. If you add the two Naira fifty kobo of this new sale, then I owe only six Naira, fifty kobo instead of a big amount. (Edodi leaves and after a few minutes reappears with the two-Naira-fifty-kobo bottle of ofofop.)

This good-natured encounter is a typical transaction among the Biase. Although it is the seller, Edodi, who loses out in what she describes as a charity business, she still finds an explanation for why people like Obazi are always in debt: "In this small place, we don't have the things you people in Calabar and Lagos have. So our men have to drink in the evenings and converse with their friends instead of going to a cinema or hotel as people in the city do. I am not making any money from the drinks, but if I buy sixty Naira's worth of ofofop, I make sure I get that sixty Naira back. No problem if they owe me only the profit.

To compound the problem for traders, none of them admitted keeping a written record of debtors. Some have never learned to read and write; one said he always loses papers he keeps around the house and does not consider written records the safest way to keep track of debtors. The result is that the names of debtors and the amounts they owe are sometimes forgotten and written off.

Another trader who was forced by his experience to emigrate from Agwagune felt less charitable:

> Our people are not honest. When you try to help them, you do not know that they are planning to make you useless. I always tried to help people by selling my things on credit. But they always waited for me to ask them for my money before they promised to pay. If I don't ask them, the money is gone. When I remember and ask, he starts to argue. So one day I asked, why am I doing this? I sold all the remaining goods in my store and used the money to go to Kano to look for work.

The experiences reported so far have been explained as inevitable under a market exchange system that operates without conflict. Meneses (1987) reminds us of Foster's earlier argument (1978) that trading is essentially conflictual and thrives successfully in an atmosphere where people are "freed from the usual obligations of generosity or kindness" (1978: 233). In such a relationship, both buyer and seller are able to manipulate the transaction to personal advantage, but the advantage is greater if the participants are members of different ethnic groups or social classes. Judging from the cases reported above, Biase market participants expect socioeconomic equals to feel social obligations, which contributes to our lack of success in the market enterprise.

These reported cases, however, fuel the impression that putting us into the market economy will affect us adversely and make it difficult for us to save. The general picture is different from these extreme cases of debt because most of us pay for what we buy; besides, such adverse effects may be offset if economic growth is stimulated. Moreover, the prospect of savings is only hypothetical because only a few of us are able to save. The few who save do not use their savings to invest in the agricultural economy but invest outside the rural community or, according to Haugerud, reinvest in a variety of businesses other than farming (1989). This would be changed under daily marketing as supporting industries developed; it will become more convenient for some traders and other visitors to live within the rural villages, thus furthering opportunities for economic growth.

Abini is currently experiencing a partial confirmation of this suggestion in the growth of what the people describe as an evening market. The market is said to have arisen from the need to satisfy the demands of migrant workers who return from the bush and have no place of their own to cook and eat. When they returned, they would go around the town calling out their needs so people who had those things would sell them to them. Eventually some Abini women started taking food to the village path that is also the main route through the community in order to make it easy for the migrant workers to buy food. As time went on, sellers of nonfood items started taking their goods to the places, which eventually developed into a minimarket. This evening market has become the busiest place in town and is making the town's night life last longer than it used to. Since Abini is located on a major highway, it is possible that as trade increases in this new market, more travelers will stop over to purchase bulk commodities, which may in turn stimulate production and the community's economy.

The opinion that daily markets will adversely affect rural communities assumes that only those within the communities will be involved in daily marketing. On the contrary, people in the community will still live as well as they do now if daily markets succeed because the increased demand will better absorb the current surplus which is largely wasted. Besides, there are only a few villages today that do not combine the production of subsistence crops with the production of cash crops and market entrepreneurship (Turrittin 1988). The idea that periodic markets are best for rural peoples, since they do not produce enough to support a daily market, assumes that markets function only for the trading of items produced within the market's location. There is hardly any community operating a daily market that depends on only the products of its members.

And the market system has a contagious effect that draws seemingly unenterprising members of the society into the dynamics of the system. This dynamics operates as a *reciprocal patronage strategy* between buyers and sellers as evidenced by *traffic light hawkers,* or street traders. Travelers through some African cities notice mobile markets along traffic routes. Young men hawking no more than two or three items run alongside cars, banging on car windows and pleading for patronage. This eventually leads to the appearance of more hawkers and the increasing patronage of the drivers.

If you drive through rural villages located on Nigerian highways you notice roadside minimarketing. Villagers line up along the road daily, selling a variety of food items, including plantains, yams, fish, and gari. Rural people want to come out and join the rest of the world. I believe the world should listen to what we are saying instead of pushing us back into our rural "sanctuary." The relationship we have to our market system and the conditions that affect the movement of persons, goods, and services are relevant to understanding our present economic circumstances. The extent to which rural economic growth can be sustained by factors such as passable roads and a market depends on how other aspects of our cultural systems operate with few constraints from the requirements of tradition and paternalistic theorizing by strangers.

Privatization and the "Discovery" of Resources

Increased exposure to the market has no doubt affected the egalitarian structure of the rural economy by adjusting the terms of common property ownership. This seems to confirm the fear that privatization would replace communal ownership as a bad consequence of involving rural peoples in a market economy. But as evidenced by numerous ethnographic reports, there are structural inequalities existing within even the most egalitarian types of societies; privatization has rarely been absent even among people such as the !Kung, who vigorously asserted ownership of waterholes against neighboring bands (Lee 1984). This principle was based on a tradition that accords political and territorial priority to the earliest occupiers of a territory, so that subsequent settlers defer to indigenes in matters of equality. As the Biase case indicates, even this minimal privatization has increased in recent times with the consolidation of vital resources in the hands of a few lineage men.

I noted earlier that although Biase land belongs to patrilineages, a strategy of resource "discovery" allows some lineage members to exercise exclusive rights over some land resources they find in the bush. Such finds are reported to lineages, who approve rights of exclusive ownership to the finder. Consequently, some fishponds and trees are owned by individuals whose only right to ownership is lineage sanction of a chance discovery. There are trees in the Agwagune forest that belong to specific individuals who have the right to cut them down and use them for their needs. While working in the forests, Agwagune men locate fruit trees, naturally occurring fishponds, and commercial trees in their lineage forests. They claim ownership of a tree by simply cutting a mark on it so that whoever else sees the cut mark knows that the tree has already been claimed and will not attempt to claim ownership. When the tree matures enough to give good lumber and if the owner needs to cut it, he meets the lineage, offers them a bottle of wine, and gets their permission to cut the tree. Very often he takes relatives or friends to the spot and shows them the tree so his claim can be corroborated by a third party; this is helpful because, as was noted earlier, tree ownership is occasionally disputed. In recent years, the need for private ownership of trees has increased because of the demand for lumber for constructing permanent houses. But because of the high cost of building materials, tree owners who cannot afford the additional cost of house construction have left their trees standing. Should a person die before his tree is ready for cutting, ownership automatically reverts to members of his household. To decrease forest depletion due to the ease of moving lumber out of Abini, the Abini do not practice individual tree ownership. Anyone needing lumber informs lineage members, who arrange to sell him a tree.

Since most forest exploitation rights are vested in the patrilineages, there is no standard rule on the use of the resources in our villages. Different lineages operate within their own rules, which may vary from or contradict the rules of other lineages. In Ase Egwa Imo lineage members can own as many trees as possible without facing any problem from the lineage. When a man wants to fell a tree, he offers a bottle of wine to the lineage before he begins work. In Igbadara, where there is less land, no one is permitted to own more than two trees; and anyone who marks a tree must inform the lineage immediately by paying two Naira forty kobo or one Naira twenty kobo, depending on the category of the tree. After this payment, the owner no longer needs to inform the lineage before felling the tree. The Igbadara rule allowing ownership of only two trees has less to do with controlling depletion of forest resources than with ensuring that every man has an opportunity to own a tree. In one of the lineage meetings, a man who was reported to have marked about six trees in various parts of Igbadara forest had to transfer ownership of four of them to other lineage members, who paid him a token one Naira for each tree.

As will be discussed later, gender classificatory practices that result in stratification also affect traditional laws resulting in privatization of resources. For instance, women do not generally own trees. The reason some informants gave is that women are not responsible for building the family houses for which the lumber is needed. Women also do not need to sell trees because such work is only for men. It has therefore become the rule that women do not mark or own trees. Among the women I spoke with, there was a general feeling that this was not a problem; some women said they have never thought about why only men can own trees, but that they would have little problem getting lumber if they needed it for house repairs. A more logical reason for the exclusion of women may be that women rarely go deep into the forest where men usually find the trees they mark. Men go the forest to hunt, to cut stakes for yam shoots, to find sticks for making traps, to clear forest for farming, and to do other things that take them much deeper into the secondary forests than women. Activities such as picking bush mango take women into the forest, but they rarely admit marking a tree. A woman who admitted doing so said she told her sons who claimed ownership of the tree. That way she did not have to explain to the lineage why she had marked a tree. A positive effect of this system of tree ownership among the Agwagune is that trees that would ordinarily be randomly felled are protected for many years; it also limits the number of trees individuals are permitted to own, regulates access to trees, and cuts down the incidence of tree depletion.

Some of the fishing methods the Agwagune use also involve privatization. There are three types of *bob* (fishponds) in Odumugom-Agwagune. Two types occur naturally, when water is trapped in large ponds along with a fair amount of fish as floodwater recedes. Such ponds are owned by either lineages or the entire village, and everyone has equal access to them when communal fishing is announced.

But one type of bob that promotes privatization is the *egimi*, which is man-made. In the dry season some men identify an area expected to flood during the rains and dig it to a depth of about eight feet and a width of about fifteen to twenty feet. When the bush floods in the rainy season, fish and water are trapped in the ponds. Between March and April, when the bush is sufficiently dry, Egimi owners hire workers to empty the pond of water. About five to ten men go down into the pond with large bowls and scoop water from the pond until there is only a foot of water left. The fish left behind are then simply picked up by hand or speared. This type of pond fishing is more popular with Itu-Agwagune fishermen, who supply almost 50 percent of the fish the Agwagune sell.

On the death of the owner of an egimi, the egimi passes on to the males in his patrilineage. In Itu-Agwagune there are two households that have more egimi than anyone else—fifteen for one and twelve for the other. One of the owners had dug up two by himself, and the other inherited all his egimi from his father. A group of two to five men may own an egimi together to lessen the work involved in maintaining it; but only about 2 percent of Agwagune men own egimi. There is no restriction on the number of egimi an individual may own, but only a few of them are profitable, since very few fish are trapped in the majority of egimi. Some egimi owners may find as much as one hundred pounds of fish in one egimi and virtually nothing in another; such uncertainty discourages most men from this venture. However, one lucky catch may be enough to change one's economic position.

Some lineages own more resources than others and use them to improve their socioeconomic condition through a system that recruits labor from other lineages. Bob ownership demonstrates this form of privatization. Naturally occurring fish ponds in lineage bush customarily belong to patrilineages, who periodically inspect them to determine when to announce that people should go there and fish. On the appointed date, everyone gathers at the pond in preparation for the event. Women usually make up most of the crowd because most of the fishponds are said to contain only small fishes; the men usually stay away from such ponds. They prefer the few ponds with bigger fish, where they can use spears and machetes. Before fishing begins, the lineage designates a woman who ritualistically steps into the water with her hoopnet, prays for fishing to be good, and dips her net into the water; then everyone else follows. As people fish they occasionally come out of the pond to empty the contents of their nets into containers. After about three hours of fishing, people start climbing out of the pond and go to the fish tax area where some of their fish is taxed or taken by lineage representatives.

While fishing is going on, men of the patrilineage are busy erecting the tax booth, *ediba bob,* a few yards away, by the only exit from the pond. As people return with their catch, they put their fish containers down. The men then take whatever amount of fish they wish from the catch before the person leaves. Members of the lineage that owns the pond and the community's senior citizens (or as is often the case, their designated representatives) are customarily exempted from the tax. Older members too infirm to participate send representatives. When the representatives get to ediba bob, they mention the name of the person they represent and are allowed to keep all their fish. A few people have attempted to use the concession to avoid being taxed. On one occasion three people gave the name of the same senior citizen that they claimed to represent; but such cases are easily detected.

The ediba bob is occasionally a place of anger. Out of about 150 women who fished on one day, 48 were not taxed because they represented senior citizens. Of those taxed, 87 people argued bitterly with the men, and 3 angrily abandoned the rest of their catch. It is against the rule of the Agwagune for people to take back the fish that were taken out of their containers; some women, anticipating a large tax, quickly grab their containers after the first fish is taken from them and run off without further penalty. After everyone's fish are taxed, ediba bob men take a portion which they share on the spot among themselves, then take the rest to the home of the lineage onun where the fish are either shared among all the households (plate 16) or sold collectively by the lineage in order to get cash to meet lineage obligations. Commonly the lineage ends up with about 50 percent of the total catch from the fishing. In a sense, the lineage employs the community to work for it and pays members about 40 percent of the day's total earnings in the form of the fish people are allowed to take away. Through this system, lineages benefit by using the labor supply of other lineages while expending little of their own. This is particularly beneficial for lineages with an insufficient labor supply.

Our men enjoy organizing bob fishing because it provides an opportunity for them to get together and collect taxes that will take care of some lineage obligations. Collecting the fish tax does not involve any extensive organization, and is a privilege we are willing to participate in, particularly because of individual rewards like an extra amount of fish distributed among the fish tax collectors at the end of their task.

Other privately owned resources include bush-mango, palm trees, and thatch roofing materials (plate 23), which are all subject to various use regulations determined by their owners. People are not hesitant to accumulate resources and make more income than their neighbors. A specific instance of this was an incident in which one fish seller misled another. Adiaha, a fish trader from Egbisim, came to Emomoro in the morning to buy fish from the men who often returned from checking their fishing shortly before 9 A.M. On this day,the men were late in returning. Meanwhile Uka, another buyer living in Emomoro, had been awaiting the men's return. When Uka saw Adiaha, he intercepted her and told her that the Emomoro men were not allowed to go fishing that day because the Onun had announced a communal work project to which all of them had gone. Adiaha went back to Egbisim feeling disappointed. When the men returned a short time later, Uka bought up most of the fish on credit, sold it, and made a substantial profit. Adiaha was not happy with Uka when he appeared in Egbisim selling the fish he had told her would not be there, but she did not condemn him. She made a joke of the incident and said Uka was a smooth trickster who knew how to survive. When the news spread around the village, no one regarded Uka's deception as wrong; a few people commended him for being "city smart."[2]

Privatization has also affected the resources available for doing community projects. When Emomoro/Egbisim needed to build a town hall, no individuals or lineages offered to donate any trees. Since many trees are not collectively owned, it was impossible to convince individual owners to give up their trees even for the benefit of the entire community. The villages ended up buying their own forest trees from individuals in order to carry out the communal project. On another oc-

2. This trend toward privatization has also been reported for other rural societies, some of which have adopted new social systems that actually foster private accumulation and inequality. When the Kikuyu moved from a family system based on the lineage to one based on households, the common rights all households had to material resources reverted to control by any one of them; rather than promoting the sharing of wealth, the new system became a mechanism for accumulation and inequality (Bates 1990).

casion, when one of the villages needed a ram for a sacrifice, the one best suited for the purpose belonged to one of the Inun, who sold it to the village for about double the going prices.

These circumstances have not caused absolute differences in the socioeconomic circumstances of the Biase, but the changes recorded have been significant in relative terms. In some villages up to 90 percent of houses are still made of the traditional wattle and daub with thatched roofs, but there has been an increase in the rate of replacement with permanent concrete and corrugated iron sheets (plate 24). In the year of my study, two of four houses that had their thatch roofing changed belonged to Egimi owners. The other two belonged to tree owners who hired lumberjacks to cut down for them the trees which they used to remodel their houses. The new look some houses are acquiring has resulted in the redesign and rearrangement of the decor of rural houses as the earth mounds traditionally used as sleeping and seating areas are being replaced by wooden beds, tables, and chairs. These modest changes are strong indicators of rural socioeconomic differences, because the high cost of iron sheeting limits its use to only a few people. It is also a classificatory index we use to assess one another's economic status. When Asibong changed the roof on his family house, his friends said things were "good" for him, and that he would no longer suffer a leaking roof during the rains like the rest of them.

Conclusion

Among the Biase, the rural socioeconomic homogeneity which is often attributed to the existence of leveling mechanisms (Nash 1971) that keep rural people fairly equal in wealth or in a situation of shared poverty (Huizer 1970) seems to be disappearing rapidly.

Although our people are aware of their problem of economic growth, the institutional support mechanisms currently in place are inadequate for solving these problems. Even as the Agwagune seek economic development, their desire to reassert their historical supremacy makes them inflexible to the new economic realities they face and the way those realities are shaped by gender, age, and new political relations. In the next chapter I will explore how gender relations have complicated this quest and reconfigured labor relations.

SIX

Biase Social Organization and the Reconstruction of Gender Roles

> In 1959 in Tanzania while I was studying the people of a village of the Wahi in the country of the Turu, I was startled one day by the refusal of a group of brothers to help their ancient mother cultivate her field. The old lady, Nyankambi . . . was so agitated by the danger of permanent damage to her crops that she prevailed upon me to carry to a mill five miles away a load of grain she wanted ground. With the flour she planned to make some beer with which to pay her sons and others to help her cultivate her field. This incident and others of the same type raised questions in my mind about the theory I was working with—a theory based on the assumption of cooperation and mutual aid among the members of the family and the larger community.
>
> (Harold Schneider, *The Wahi Wanyaturu: Economics in an African Society* [1970]: 1; Cited in Donham 1985: 11)

The most popular idea about kinship relations in Africa is also the major misconception about it. This is the notion that blood is thicker than water, which is no longer merely a romantic myth chronicled in the works of some social anthropologists but an important concept whose preeminence all anthropologists still feel duty bound to prove. Kinship and its functions are seen as the dominant principles that determine every facet of the lives of those of us who live in what are described as headless societies in the classical social science literature. In short, an understanding of the configuration and functions of kinship relations in Africa is synonymous with an understanding of every facet of African social life. So the notion that blood is thicker than water becomes important in sanctioning the obligations of kinship and serves as an easy guide for decoding and interpreting our actions. "There is the temptation to assume that kinship linkages are all-important and that once one has demonstrated that members of a particular group are related to one another, one has laid bare the group's structure" (cited in Boyle 1977). Consequently little attention is paid to the more significant social institutions and behavioral precepts within which the kinship principle operates. Other institutions, such as men's and women's associations, age sets, work groups, and thrift associations have over the years been more important to individuals in a society than their families, not only because some of these organizations are more stable than family groups but also because they are more relevant to the individual in sustaining and preserving life, promoting good health, and nourishing economic and social relations.

Vaughan has observed that the African family "does not constitute an extensive group of adults intrinsically pledged to mutual support and security" (1986: 169). Indeed, the real situation is that social and political power are important goals and many of our men and women realize that the attainment of that goal awaits them "not in family ties but in impersonal contacts with economic and political entrepreneurs" (1992: 52).

In Biase, as in other technologically less-complex cultures, kinship relations significantly affect other forms of social interaction but at the same time appear to be a less profound basis for social organization than nondomestic social groups. Such mystical associations as Inyono and Ekpe and such men's secular associations as Abu and Ebrambi performed functions of social bonding that in a number of circumstances were more important for the individual than kinship in shaping, ordering, and regulating the behavior of people.

The diminishing role of kinship relations in cooperative behavior is more apparent in technologically complex cultures, where social relations increasingly coalesce around nondomestic units (Harris 1981), than in technologically less complex cultures. But the Biase situation suggests that the role of kinship is also diminished in less complex societies because successive interactions with external institutions change rural socioeconomic relationships and lower the effectiveness of kinship relations. No doubt ties of kinship are important in determining relatedness, but their functions do not extend much beyond locating the individual on the kinship map. In Biase, men's social groups historically made decisions that affected lineages, families, and individuals and in the process generated the respect and fear that were necessary for social cohesion. In this chapter I will examine how the loss of authority by some of these groups deflected authority to other groups, restructured Biase social organization, shaped its economy, and exerted a significant influence on gender relations.

Biase Mystical and Secular Associations

In Biase we believe there are mystical associations, such as Inyono, that incorporate practices which are beyond the knowledge and awareness of most community members. Although we are unable to prove the existence of Inyono, we are convinced that its performances rely on magical rituals that sometimes involve cabalistic sacrifice of close family members. We also believe these organizations limit membership to persons willing to fulfill specific mysterious obligations as a way of maintaining their membership in Inyono. The rewards for members may include greater material wealth or a longer life. Although no one in our villages admits membership in Inyono, every villager seems to know who the members are.

Secular associations differ from the mystical ones in their openness: there is little linkage with mysticism, membership is ascribed on the basis of birth, a public initiation, or gender, and a public performance is important for announcing the attainment of membership. The women's associations, such as Egup, which incorporate both mystical and secular elements, lie midway between the men's secular and mystical associations. Membership in Egup is by birth and gender. Although there is no public initiation, there is an elaborate public performance. A priestess who is possessed by a water spirit performs the Egup rituals at night. She lives in the Egup shrine and is rarely seen by community members. Egup rituals are said to consist of mystical elements known only to a few older Egup members. The mystery surrounding these groups has had significant consequences for other patterns of social interaction.

Many years ago, families with members of feared mystical associations were venerated because some of their members possessed certain mystical abilities. They were believed to be able to cause various degrees of harm to anyone they chose, especially family members who were the target of cabalistic sacrifices to Inyono demands. Because of a circle of allegiance that provided primarily for Inyono members within and outside the community, individuals were able to consolidate power and prestige which provided a way to advance political interests and maximize economic goals. This had some economic and social consequences for our villages.

The social consequence is that kinship relations, while manifest in the existence of structured kin groups, only tended to camouflage a subset of values and codes beyond the control or understanding of most kin group members, but which functioned to hold related persons together. There is a song of the Lozi of Zambia which explains this paradox of kinship. In his article "African Systems of Thought," Ivan Karp records the lyrics of this song: "He who kills me, who will it be but my kinsman?/He who succours me, who will it be but my kinsman?" (1986: 208). A younger informant told a story of her father's involvement in Inyono. Five years before my study, Inyono had informed her father, who was a member, that it was his turn to offer a member of his family to Inyono and that they specifically mentioned her as the desired offering. When her father refused to make the offering, Inyono took him in her stead. Such stories are common in Biase villages. Some even narrated tales of Inyono members who choked and died trying to "eat" them. No one has ever admitted being a member of the Inyono although specific names often came up in conversations. In Agwagune I spoke with two suspected Inyono members who said there were no longer Inyono members in our villages. One said, the spiritualist church "Brotherhood has cleansed this place of such evil."

Economically, community elders belonging to mystical associations had little problem getting youths to fill labor needs such as helping on the farms or taking yam canoes up the Cross River. Youths aspiring to economic success always sought to associate with those elders rumored to be higher up in the hierarchy of the mystical associations, because being successful while young without the support of an elderly person was perceived as an affront to the authority of elders. Young men eagerly accompanied community elders on long trading trips that kept them away from the village for as long as three months. A few of the youths who benefited from learning to trade later established their own businesses along the upper Cross River away from our villages and the scrutiny of the elders.

The general impression today is that our mystical associations are no longer as powerful as they once were. In the early 1970s a group of young men from my village traveled north of Biase to purchase *igwu*, a protective magic which was expected to kill anyone who had the intention of committing evil against another. Igwu has been performing as expected because men and women who were often mentioned in connection with mystical associations have died mysteriously, apparently killed by igwu, while others have joined spiritualist churches.

Two results are apparent. First, the Biase now have less fear of supernatural consequences from elements they do not understand. Accusations of witchcraft have diminished, because fewer people are suspected of engaging in practices that would expose them to the wrath of igwu. Loss of fear among the young has minimized the adherence to social decorum and reverence toward community elders which seemed to be important for compelling conformity to rules and regulations. An economic consequence of this diminished institutional authority is an effect on labor relations. An increasing number of Biase youths feel free to redirect their labor to private ventures without fearing for their lives, and more experienced traders cannot get the labor they need to continue or to expand their enterprises. Furthermore, since mystical associations were important for ordering kinship relations their loss of authority weakened kinship ties and left little basis for the consolidation or strengthening of ties. People talked openly about dying Inyono members who publicly confessed their responsibility for the deaths of family members over a number of years.

The Biase are aware of their relatedness and can easily make connections from every lineage member to an apical ancestor. But finding a place on the kinship map is only a way of placing a person within the social order rather than a primary basis for social cooperation or mobility; familial obligations wane when people have concrete goals. People in my village often defend their relatives verbally but become antagonistic to them when they want material help.

Since people live together according to descent, kinship might seem to have more of an impact than I suggest. But it does not. Some people who trace relationships to relatively well-off people are themselves less well-off than most people in the village. Some talked about relatively wealthy family members living in the village or outside Biase who rarely provide help. Such circumstances result in numerous membership shifts between lineages. One reason people give for resigning from their lineage to establish membership in another is that they are not being treated fairly. One man who was involved in a dispute with his paternal uncle moved out of his patrilineage compound into a house in his matrilineage compound after the dispute was settled in favor of his uncle. Resigning from one's lineage is done informally and without consultation with one's lineage or family. An individual wanting to resign from one lineage to join another simply presents drinks to the Onun and elders of the prospective new lineage and informs them about his or her intention.

There were old men and women living in houses with leaky roofs and broken walls while their sons and daughters next door seemed not to notice. Egana was a handicapped forty-eight-year-old man (blind and with a leg deformity) who lived alone in a dilapidated house that had been abandoned by its previous inhabitants. Although there were a total of fourteen working adults in his extended family, no one offered him a place to live and he always depended on the goodwill of other village people for water, food, and clothes. Members of his family, including his mother, his three sisters, a brother, and several cousins, said they shunned him because he was an ungrateful fellow. An example often given of his ingratitude was that he was overly reserved in expressing appreciation for the occasional favors people showed him. As one of the foremost drummers in the village, Egana associated more with members of his age sets and his friends. To help him make a living, some of his friends and members of his age set would give him their fish to sell and specify the amount they expected back. They often expected him to sell the fish at a price a good deal higher than their suggested price (which he usually did) and keep the profit for himself. On one occasion, Egana got into trouble after selling fish for a friend because he did not return the money. In the confrontation that later ensued, not only people from his immediate family but members of his patrilineage and matrilineage came to his defense. One of those who had lost money derided Egana's relatives: "I never knew Egana still had this many people in his family. Where were you all when he was dying of hunger?"

On another occasion members of my patrilineage refused to contribute money for taking our ill lineage Onun to a hospital. One woman complained that she had been left on her own to care for her mother who had been ill for many years without any help from lineage members. She asked the gathering who among us was not aware of her mother's sickness and which of us had ever offered to pay for her medication. Without waiting for an answer, she walked out of the meeting.

In Agwagune, there seemed to be little cooperation in work even among related households (table 9), although 66 percent of the households I surveyed said they got extra labor from their kin. People tended to base their answers on what my culture classifies as culturally appropriate, since the family is indeed the first place we turn to for labor. But since all related adults own their own farms, it is difficult to get them to defer their own work in order to help family members. A more critical reason was that nonfamilial relationships provide more profound bonding for individuals (plate 20). Friendship groups or age sets were the basis for cooperative behavior: traveling together, fishing, organizing work, and expressing goodwill or assigning trust.

Such cooperation tends to center more on external than domestic relations, decreases the importance of the family as the primary unit of social organization, diminishes traditional male authority, and opens the way for women and young men to contest the traditional social polity and play more visible roles in reshaping traditional institutions.

TABLE 9 Frequency of Labor Exchange in Five Sets of Related Emomoro Households

From	To Household										
Household	21	26	25	36	122	126	47	110	99	133	134
21		0	1	0	0	0	0	0	0	1	2
26	0		0	0	3	0	0	0	0	0	0
25	0	0		0	0	0	0	0	0	0	0
36	0	0	2		0	2	0	2	0	0	0
122	0	3	1	0		0	0	0	0	0	0
126	0	0	0	2	0		0	2	0	0	0
47	0	0	0	0	0	0		0	3	0	0
110	0	0	0	2	0	2	0		0	0	0
99	0	0	0	0	0	0	2	0		0	0
133	1	0	0	0	0	0	0	0	0		0
134	2	0	0	0	0	0	0	0	0	0	

Source: Adapted from table 5.1 in Allen W. Johnson, *Quantification in Cultural Anthropology: An Introduction to Research Design* (1978), p. 100. Stanford: Stanford University Press.

These changed circumstances affect social relationships in restructuring gender-based roles that are critical for organizing society. The process of this status acquisition and the evolution of gender roles in the management of the Biase economy will be the focus of the rest of this chapter.

Social Construction of Status and Gender Roles

The earlier tendency of social scientists to perceive rural societies as homogeneous and egalitarian made rural-based inequality difficult to understand beyond the familiar assumption that women are generally in a subordinate status. Social scientists are now increasingly aware that the daily processes of life in villages revolve around a series of gender- and age-based performances that conflict with social ideals but which are resolved within mutually negotiated codes. The end result usually changes traditionally accepted roles, reforming old actors or putting new actors in charge of how communal resources are generated, managed, and applied to general communal welfare. Community members applaud these performances by assigning authority to individuals based on the degree to which the individual contributes to socially sanctioned good.

The relative standing of an individual in a particular culture is said to have an isomorphic relationship to the individual's wealth and power. Hatch summarizes this opinion: since "individuals find wealth and power desirable, they respect or envy those who have more than they; hence, an individual's position within the economic and political order is the basis for his or her standing within the hierarchy of social honor" (1989: 342). Although Hatch disagrees with this opinion because social honor is not always isomorphic with material inequality, his earlier investigations among New Zealand and California farm communities showed wealth to be a crucial indicator of a person's social standing (1987: 48). This model is useful for understanding the effect of wealth in defining authority in apparently egalitarian cultures. Its relevance to the Biase case concerns two key suggestions we made earlier.

First, traditional authority is no longer exclusively controlled by elderly men but also by young lineage men and women. Young people's growing loss of confidence in elders results in contempt for normative rules, irreverence for fundamental beliefs and ideologies of the Biase, and disrespect for elders. Second, the resulting loss of the political and economic authority of men has opened new opportunities for Biase women. Today wage labor is an important index for assigning prestige. In my village we often say that "those whose feet are firmly on the ground have crossed the river and come back with zinc" (referring to resourceful Agwagune emigrants who have replaced their traditional thatch roofs with corrugated iron sheets). Women are profoundly involved in this transformation. They are increasing their participation in the local economy and seeking wage labor outside the community. The greater economic independence now results from their role as the main food producers, the primary entrepreneurs in Biase, and the major financers of communal projects.

Wealth seems to be important for the status of non–village dwellers and is very often assumed to be a consequence of living in the city. Our men treat women emigrants from the city with more courtesy than their counterparts in the village. Men fawn over women visiting from the cities and offer to run errands they do not often do for rural women. Biase women who hold low- and high-level government positions are treated with deference by rural men. Young men often walk four to nine kilometers to do errands for women visiting from the city.

The lower status of rural women among their men may be attributable to the fact that we see ourselves as all occupying the same socioeconomic level. Since I am familiar with the capabilities of my neighbors because of our minimal variation in occupations, I can guess my neighbor's income level. Using that criterion, you know that there is rarely anyone in the village so extraordinary that his income level commands any more respect than yours or your relatives'. There are no visible symbols of wealth that mark such differentiation, as may be the case in the cities. Elvin Hatch, for example, observes of California and New Zealand farm communities that "the essentials of the prestige hierarchy (which is a system of subjective evaluations on the part of the members of a social body) are immediately deducible from the objective, or directly observable, features of the economic order" (1987: 37). Using directly observable features such as size of property and the quality of a farmer's sheep and house, people in his study made judgments about who was wealthiest in the community. Among Biase rural people, where everyone is perceived to be on the same socioeconomic level, prestige ranking within the community is determined less by wealth than by noneconomic features. Wealth is a significant determinant of prestige only if it is acquired outside our villages. We believe that people with extraordinary skills leave the village to do the "white man's" work and are accorded recognition by the government by their earnings. In this regard, men and women working for the government or otherwise earning wages from labor outside their villages are respected much more than those toiling in the village. Whenever anyone from the city takes a bucket to the stream, a village resident offers to get the water. But rural women leaders and other elderly men and women of the community perform such tasks for themselves without offers of assistance from anyone.

One explanation given for the concessions rural people offer city women is that they are not strong enough to do the hard physical labor that rural women routinely perform (Moran 1988). Hollos (1991) and Moran suggest that such ideas also limit the participation of city women in the politics of village women's associations. That may indeed be the case if city women are clearly separated from the rural economy of their village counterparts, as when a woman's high level of education offers her the opportunity to migrate to the city for a relatively high-paying job.[1] But because of a generally low level of education, returns per unit of labor are not so much higher for wage employment that people are drawn away from farmwork (Barber 1960). Our women, like our men, alternate wage labor with farmwork. Because of her bicultural expertise in rural and city ways, the city dweller is recognized to have familiarity

1. Between 1985 and 1990, 5 percent of Agwagune primary six children went into secondary schools; in Emomoro, only one person continued to a postsecondary institution in 1990.

with both city and rural survival strategies; this additional knowledge gives her a higher status in the public sphere than her rural counterpart.[2]

Although the rural woman may make more money than her counterpart in the city, she has not elevated herself above the rural order if she still ekes out a living by working in mud, flood, and dust like everyone else. One Christmas day when some women dressed in festive clothes passed by a lineage center, a group of young men lounging at the center commented humorously that the women looked beautiful because of the fish money of their husbands. The implication is that the rural woman who exhibits economic success is only able to do so because of the benevolence of the men whose fish they sell for profit.

Economic and political powers therefore vary independently of each other when it comes to ranking women, but this is not often the case in ranking men (tables 10 and 11). I mentioned in earlier chapters that our women are more involved in on- and off-farm labor than our men, and that this gives them a relatively higher income. Their freedom to control most of their labor and income has encouraged them to form a women's association that exerts a dominant influence on Biase affairs and consistently applies rewards and punishments to its members, as well as influencing some of the decisions of their men.

But this economic strength does not translate into political authority outside the caucuses of women, although it puts them in a good bargaining position in certain matters. Consequently, although women have economic power, they lack political power, since prestige in the economic sphere is subsumed under an asymmetric traditional power structure controlled by men (table 10). This situation runs counter to what the literature suggests, because Biase women's significant contribution to the economy should be rewarded with a high ranking by men.

Numerous examples in the urban environment in both developed and developing economies indicate that contributions to the economy and control of productive resources are important determinants of the placement of women in the social hierarchy. In farming cultures like the Biase, where women's contribution to subsistence is substantial, ethnography tells us that gender inequality is narrower than in cultures where women contribute little. Other studies (Sanday 1974; Safilios-

2. Over 60 percent of Biase women living in Calabar, the state capital, maintain their own households, but over 90 percent of those with husbands earn much less than their rural counterparts.

Rothschild 1988; Knack 1989) also suggest that when significant food contributions and greater economic power combine, women should be expected not only to have optimal equality but also be able to exercise political authority almost on a par with men. As Sanday (1974) puts it, "where control and produce are linked and a competitive market exists, female power is likely to develop if females are actively involved in producing valued market goods" (cited in Knack 1989). Sanday mentions four criteria that will bring the situation about: female ability to control and allocate goods beyond the domestic sphere, recognition of and demand for women's produce, participation in political decision making affecting people outside the domestic unit, and an effective interest group representing women (Knack 1989). Knack's study of contemporary Southern Paiutes confirms Sanday's hypothesis, as do other studies, including Wilson-Moore's, which show that a loss of social standing follows women's loss of resource control (1989). Wilson-Moore reports that although the Chuchuli women of northwestern Bangladesh have

TABLE 10 Assessment of Wealth by Men and Women in Biase

Those	Rich Person		Big Farmer		Generous Person	
Ranked	M	F	M	F	M	F
M4*	20.0	20.0	9.5	10.5	17.0	19.5
F2+†	12.0	18.0	18.5	20.0	8.0	20.0
M1*	13.5	15.0	15.5	18.0	9.5	12.5
M3+	10.0	13.5	14.0	16.5	9.5	12.0
F1*	17.5	20.0	5.5	6.5	10.0	19.0
M2+	5.5	7.5	9.0	16.5	5.0	5.0

Note: M = male; F = female; * = city resident; + = village resident; † = leader of women.

TABLE 11 Assessment of Social Status by Men and Women in Biase

	Those Ranked			
	Persuasive		Respected	
	M	F	M	F
M4*	18.5	20.0	20.0	19.5
F2+†	17.5	20.0	13.5	20.0
M1*	18.5	19.5	18.5	19.5
M3+	20.0	20.0	10.0	11.0
F1*	5.0	17.5	10.5	20.0
M2+	20.0	20.0	20.0	20.0

Note: M = male; F = female; * = city resident; + = village resident; † = leader of women.

experienced economic success through managing homestead gardens, this has had no effect on their social status because the income from the sale of produce is handled by the men of the household. From the discussion in the last chapter and as will become clear from this chapter, Biase women also fulfill these four criteria; but the recognition and acceptance men show them reflects the standing of the women's corporate group rather than that of the individual woman. We appreciate our women individually for their personal achievements and their contribution to the society, but such recognition rarely raises them in the social hierarchy and may have the reverse effect of lowering their status. One example occurred soon after I arrived in Biase to begin my fieldwork.

I was directed to two women leaders whose cooperation was necessary for me to be allowed into some of the women's meetings. Although the two women often represented their villages in dealings with neighboring villages, I was cautioned to be careful in associating with them because they were "powerful" women. This is translated to mean that they have supernatural powers. One of the women had accompanied men during their search for igwu and had successfully "defeated" all the supernatural magic along the way during the quest. On another occasion she had represented the community in a land dispute that was taken before the state court and had successfully warded off the evil machinations the enemy community had directed at her. While these successes made her invaluable to the town and she was praised in songs by men and women, people also portrayed her as a witch. Some villagers advised me not to eat or drink in the woman's house for fear she would offer me as a sacrifice to strengthen her powers. However, the men with whom she had recorded those successes were locally described as "full" (that is, they were well prepared to handle any evil). So the social qualities for which men were praised were the same qualities for which women were ridiculed. But this negative portrayal of women by men is more descriptive of individual women than of the group identity (see tables 10 and 11).

While we seem not to acknowledge women's political authority, we accept and recognize that our women who have attained political authority outside the community are capable of fulfilling the responsibilities of that role. In the December 1990 local government elections conducted in Nigeria preparatory to civil rule, an Agwagune woman was one of two candidates who contested the position of councilor of the Biase local government; the male opposing candidate was also from

Agwagune. Although the woman received only a third of the total vote, 30 percent of those who voted for her were men.

This underlines the paradox in the attitudes with respect to individual versus group distinction: the public contribution of women is translatable into personal honor, but the esteem is not public; a woman is minimally effective in influencing the community. A woman is able to attain social distinction but is constrained by traditional role expectations from attaining or overtly exercising political power back in the village. The achievements of individual rural women are not accorded much importance by men, but as a group women are acknowledged by men to be the lifeblood of the community.

This is the background that informs gender relationships among the Biase and the interplay of economic and social power in the management of communal goals. While not manifestly accepting the economic power of women, our men are aware of the importance of not souring relations with them. Men admit that the women's meeting, Aka-e-mitin Aneba, is important in the economy of Biase, a point which was continuously demonstrated during the year I was there.

During the rainy season, when the last three kilometers to Egbisim village were flooded, passage was possible only by wading through the water or by using small and medium-sized dugout canoes. For three weeks schoolchildren and travelers found their way by wading neck deep in the water; visitors to the town unprepared for this adventure turned back. I learned that usually the two Inun of Emomoro/Egbisim put a canoe at this point to ferry people across, but this was not done because the Inun were still arranging to meet about the problem. The women believed that the men were unconcerned about the welfare of those needing this service since, unlike women, most men own their own canoes. They openly taunted us for being too weak to help solve the problem.

After two weeks, the local unit of Aka-e-mitin Aneba rented a medium-sized canoe from one of the men and put it at the flooded point; they then hired one of the men to operate the service for a monthly fee. Since we are traditionally in charge of making and executing social rules, they bought drinks for us and asked us to have the town crier caution the village against disorderliness at the ferry point.[3]

3. On previous occasions, canoes have capsized and valuables were lost as users struggled to get seats in the canoes. Schoolchildren were carried free of charge, commu-

Two days later, another canoe appeared at the point; but this time, it belonged to Odim, a man who charged a high fare to ferry people in his canoe. Odim refused to ferry schoolchildren because they were traditionally exempted from paying for the service. That evening, the women's meeting sent a messenger to announce in all the villages that every woman should boycott Odim's canoe; anyone who continued to use his services would pay a fine of fifty Naira. They also said a second canoe was needed to accommodate the volume of travelers and to control the rate of accidental overturning of canoes that frequently resulted from the crowd of passengers attempting to board the only available canoe. The next day, with no prior notice, the messenger went out again to announce that all women should assemble at the point from which they were going to drag a new canoe to the ferry point. The response to this call was so fast that there was no time for me to run back to my base and get my camera. Within fifteen minutes, more than two hundred women had assembled around a medium-sized canoe and proceeded to drag it about 600 yards over a steep slope to the desired crossing. A group of twenty-one men was sitting at a patrilineage center right where this event was unfolding watching the whole incident with as much fascination as the enthralled travelers who would eventually benefit from the undertaking. Some of the men seemed embarrassed just sitting there and watching women perform this onerous task, so a few left the center while some others watched through partially opened windows. The men conceded that women were better at organization and said that when men were asked to handle matters it took weeks and even months before any action was taken. Some women said it was not their job to take charge of such situations but that men always beguiled them into believing that they were better organizers.

Although the flooding is our major concern, we do not have a leadership structure that relates to it since the problem is defined more by the Biase economy than its politics. Whatever the political component is relates to our relationship with the state and our inability to attract the attention of the government. Our inadequate technology suggests we don't care about the flood problem. Flooding has simply overwhelmed our technology, because although most men own canoes, many of the canoes are old and in need of extensive repairs. Many men also fear risking the lives of their families in small canoes that are primarily built for a

nity members who were going to their farms were charged 10 kobo on their return trips, and travelers and visitors paid 50 kobo.

man and his fish. Men have constructed temporary bridges in other flooded areas and have provided private canoes in a few more, but these have little chance against the severity of the flood at certain times. Generally, the tough economic situation of our men constrains them from being more adventurous in planning projects that would minimize our flood problems.

Aka-e-mitin Aneba is important because of its strong capability to organize. In the association, women recognize and value the aggregate of material and nonmaterial achievement and qualities of fellow members such as wealth and the ability to persuade, communicate, organize, and lead. In that arena they collectively make demands, enforce rules, and punish violators of the rules. Their lack of political power in the social sphere is compensated for in an arena where they are free to exercise their power away from the influence and dictates of men. In this sphere, members accord recognition and prestige to each other's capability and honor the authority associated with it. This is particularly evident in the resolution of matters affecting association members. Disputants are summoned before the women's council, which hears a case, decides on guilt or innocence, and imposes fines. Such fines are often promptly paid or a number of restrictive sanctions are imposed on the offender. Punishments include being barred from associating with people outside the immediate household, the cessation of use rights to forest resources, or being stopped from entering the bush to do any manner of work. This is a severe penalty for people whose livelihood is based primarily on the use of forest resources and the goodwill of village people. In this way our women cohere around the variable capabilities of individual members, assign authority to individuals, and empower and strengthen their group.

The disposition of Biase women to accord prestige to qualities of their peers did not happen independently but may result from women making a living outside the rural milieu being more likely to acquire prestige among their peers than rural women. Although this no longer seems true in the women's association, it was an important inspiration for its formation. Informants said that Agwagune women had no formal association for organizing their members after the demise of Egup, which had served as the women's government. Egup was destroyed because many women were dissatisfied with the concentration of authority within a small group of Egup women who ran an undemocratic organization that gave other women little chance to express themselves. When the incumbent Onun Egup, Eba Uno Esu of Emo Edodi, died in

February 1960, the family of the woman elected to replace her protested the choice by refusing to let her serve. This protest was partly a result of fear of what is known to be the social status of Onun Egup. Once she accepts her role, the Egup queen completely gives up her relationship with her family. She never again has anything to do with her parents, children, or friends and becomes only the representative of the Egup spirit. From the moment she is captured to be queen, she loses consciousness of her humanness as she is possessed by the water spirit of Egup. The Egup queen remains in this state until her death. Many elderly community women tried unsuccessfully to organize the leaderless women around an alternative leadership until Aneji Jenny Ekeko, an Egbisim woman who operated a tailoring business outside the community in Okopedi-Itu, appeared on the scene. Having built up a great deal of respect for herself among men and women in Biase and other parts of Nigeria, Aneji had little problem organizing women around her leadership to found the Aka-e-Mitin Aneba.

With the death of the important leaders of the association, including its founder, Aneji Ekeko, in 1980, the association started experiencing difficulties which later decreased its organizational effectiveness. The most critical problem threatened the unity of the villages of Emomoro and Egbisim; for four months women of both villages traded insults and stayed away from events that traditionally required their cooperation.

The trouble came a month before my arrival in Biase. As is customary, the women of the two villages had gone to gather bush-mango in Emomoro forest; a group of Emomoro women remarked that Egbisim women had no food and that if it were not for the Emomoro fruit trees, Egbisim would starve to death. The Egbisim women regarded this statement as a serious insult and announced in the village that they would no longer participate in any social events in Emomoro. This tension caused a significant economic loss for the women of both villages because intergroup cooperation between the villages ceased, and attained a critical point between January and April which is the peak fishpond period. Pond fishing is done mainly by women and requires the participation of many people wading through the trapped water with hoop nets if any substantial amount of fish are to be caught (plate 15). Because of the problem between the villages, fewer than 50 percent of the effective number of people regularly participated, so very few fish were caught. In the past, optimum participation often resulted in an average of twenty pounds of fish per person; during the standoff, villagers caught an average of eight pounds per visit. Women reported an average income of about twenty Naira every week from the sale of fish between January and April as opposed to fifty Naira for the same period in 1989.

The economic loss was even greater for Egbisim women. Emomoro had more fishponds, which were also said to contain more fish than the Egbisim ponds. Since Egbisim women did not participate in fishing Emomoro ponds, they made very little income from selling fish; but Emomoro women also experienced a drop in income from the sale of fish since they caught fewer fish than in previous seasons. For a few weeks, it seemed as though the situation would worsen unless prompt action was taken to settle the matter so that Egbisim women could make a little money before the fishing season ended at about the end of April. Emomoro women leaders met to discuss the matter and felt that if the misunderstanding was allowed to continue beyond the fishing season and settled only after it, Egbisim women would feel that the intention was to exclude them from fishing Emomoro ponds and would not be enthusiastic about settling the dispute. At about the middle of March, representatives of the Emomoro women went to Egbisim and apologized for the insult a day before they were to fish the best pond in the village. The next day about 209 women descended into the pond and caught so many fish that some women had to get help to carry their fish home. The smallest catch I weighed was 22 pounds and belonged to a woman who left the pond early because she had been injured by the fin of a fish. On the average, individuals caught about 35 pounds of fish, which after being dried, brought up to 100 Naira in the neighboring markets.

Because of the loss of cooperation by women, lineages that owned fishponds did not check them to see how fast their ponds were drying in order to fix a date for fishing as they customarily do. When some lineages eventually visited their ponds, they found they contained only bones of dead fish all over the mud. This resulted in a loss of income for a few lineages that relied on fish tax income collected from non–lineage members to pay lineage debts or accomplish other lineage projects. Both villages lost substantial amounts of money from sales they would have made in the neighboring markets.

Relations between the two villages were further strained because at that time the men of Emomoro and Egbisim villages were also involved in a dispute of their own and were not cooperating with each other. The problem arose over the matter of ceremonial observances in memory of the deceased Onun Abu of Egbisim. Egbisim said Emomoro men were late in honoring the invitation to participate in the funerary rites of the Onun, and they imposed fines on all Emomoro men who were either late or absent. Emomoro protested the fines and walked out of the Egbisim ceremonies. Emomoro then accused Egbisim of leasing some of its village land to wine tappers from Ikom without consulting them or sharing the proceeds. Since then both villages have observed events separately which traditionally they celebrated jointly.

On two occasions, Uru Iyam, a teacher from Emomoro, invited both villages to a meeting about the matter, but the meeting never held because the Egbisim men objected to the venue. They preferred to meet at the home of the Egbisim Onun Abu, where the original disagreement occurred. Another meeting was later successfully scheduled to be held in the market square, but there were further disagreements that led to the abandonment of the talks. For over four months, both villages lived like enemies. Each village prohibited people from participating in events held in the other village. In order to avoid the escalation of hostilities, people seldom visited friends and family in the feuding villages. The Inun of the two villages could not convene a meeting of our men to discuss the issue because the men central to the dispute were members of Abu, which believes the village does not have the right to discuss Abu matters. It was only after the Inun of all Agwagune villages met and pleaded with both sides that the matter was resolved.

Certain aspects of these disputes clarify the gender dynamics in Biase and suggest a number of implications for the idea that indigenous institutions have become weaker. The demise of the women's Egup was followed by the founding of the Aka-e-Mitin Aneba by women who gained the respect of its members and were given authority to direct the agenda of the association. Consequently when operational problems arose, there was already a framework for resolving the problems and prevented them from escalating beyond the capacity of the association to resolve them. Besides, even when there was conflict women continued to be on good terms, particularly on market days, thus lessening the impact of animosity arising from the situation.

Men faced a different set of problems, which were more intractable because of their ideological origin. When the villages attempted to resolve the misunderstanding between Egbisim and Emomoro, core Abu members from both villages refused to discuss the matter because they felt the villages had no business dabbling in the internal affairs of Abu. Traditionally Abu is regarded as the principal government of the Agwagune with power to make rules and effect discipline. But like other traditional associations that had power to manage communal affairs,

Abu has lost most of its power and no longer incites the fear and reverence it used to command. Yet the pride our men have in Abu has discouraged them from organizing an alternative institution to strengthen the managerial capacity of men. Consequently, men were unable to confront and solve organizational problems before they grew too complex to be resolved by traditional processes. The results were that the conflict went on for two months longer than the women's conflict, cessation of cooperation extended to other spheres, and an external body had to intercede before the conflict could be resolved.

Gender Differences in Work and Leisure

Since the men in this study have continued to leave the economic and communal tasks to women, they might be expected to have a considerable amount of time to invest in nonfarm labor or even in working outside the villages. But the Biase, particularly the coastal communities, engage in little labor outside of farming or fishing that might absorb their leisure time. Our men therefore spend an enormous amount of time just sitting around their lineage centers.

Hill et al. (1985) observe that variability exists within communities in the amount of leisure time people have; the differences may be determined by gender (Chick 1986 a and b). On fourteen sample days in the rainy season month of July, when women were mostly involved with weeding and men fished, I observed the time when four men and four women left home and what time they returned. Men left an average of twenty-nine minutes before women and returned an average of five hours earlier. Men left again after five hours rest and returned at about the same time as women. So during this period women worked an average of five hours a day longer than men on tasks no less demanding than those of men.

Although the same trend has been reported for other societies (Levine 1988; Caplan 1989), this leisure time, which is mainly for men, is sometimes said to be beneficial to the community because of the need to rest from the long distances rural people have to walk, the greater energy requirements of rural labor, and even the need to improve social relationships by sitting around and discussing matters affecting the village or its neighbors (plate 21). But there is little understanding of why in communities where women have substantial political power and contribute significantly to village affairs, men should have most of the leisure time. Anthropologists such as Boas and Herskovits attributed leisure time to a surplus in the food supply which freed some community members to engage in specialized activities rather than for generalized indolence. Although this perspective had long been abandoned, an attempt was made to revive it at the beginning of the 1980s (Shivers 1981), but it met with strong criticism from scholars such as Gary Chick (1984, 1986 a and b). Chick stressed that leisure time is unrelated to productive activities because men who have much leisure time are minimally involved in subsistence.

According to my time-allocation survey, men spent an average of nineteen daylight hours per week just sitting at the village square, but women had less than four hours of leisure time per week. Women's "leisure" time was used for child care or braiding the hair of children and friends; sometimes they prepared meals simultaneously with these activities. Men often gathered in lineage centers, which some of them have dubbed "Radio Emo-Are", (Radio they said). Explaining the term, one of the men said, "It is at the lineage center that you first receive news about any event happening in London before the news even gets to Calabar (the Cross River State capital). If you want to know what the Nigerian president ate this morning, it is at the center you learn it. Here we have people who have every kind of information on any part of the world, although they have never stepped out of this village since their mothers gave birth to them." But rural people glorify their idle time much less than the literature suggests, which is why Green (1989) refers to it as "undesired leisure time." For instance, following the failed April 1990 coup against the government of Babangida, some village versions of the coup were so spurious that one of the men said, "If I had money to start a business venture that would take me away from this village, I would not be here listening to all this talk from people who do not even own radios." On another occasion one man rose and headed toward his house saying he was "tired of hearing all these lies from people who have never seen a newspaper."

Despite this suggestion of the lack of substance in men's use of leisure time, there were occasions when men used leisure time to exchange occupational information such as how to confront a bush pig if it was caught in a snare, where fish was likely to be more abundant at particular times of the season, or how to make palm wine better for customers. But such information sharing does not justify the incidence of idleness because successful interpersonal communication among rural Biase people does not depend upon idleness.

Men I spoke with regarding their idle time said they do more energy-consuming tasks than women and so need time to rest. They admitted that women worked longer hours than men, but said that such work was not as energy-demanding as men's: during planting season, men clear the bush, till the ground to make heaps for planting crops, cut sticks for stakes, and pin the stakes in the ground (plate 12). All the men I spoke with said staking is the most demanding part of farm work. When their yams are safely in the ground, men consider they have done their main work until harvest time.

Another way men justify their leisure time is in *igot* (village watch), in which sets take turns daily keeping watch over the village while most people are away in the bush. This is particularly important in the peak working season of January and June, when over 70 percent of the homes remain locked from early in the morning till late in the evening. In many lineage compounds, only one or two homes with elderly women remain open. Igot is necessary in the dry season because at this time cars are sometimes driven into or close to some of the villages. There have been cases of kidnapped children in some of the neighboring villages. However, igot requires only the presence of adults in the village patrolling from one lineage center to another and making sure there are no fights or arguments that may result in fighting. If there is a disturbance anywhere, the man hurries to the scene and fines those responsible for disturbing the igot. In the evening, igot men share the proceeds from their fines, usually by buying palm wine and discussing the day's successes.

Since igot did not seem to involve any more than women are capable of doing, I tried to find out why it is done only by men of the age sets. In the case of Abini village, it is explained by a system of gender separation in age-set membership. In Agwagune, where males and females belong to the same age set, women still do not participate in the village watch. Some women found it ridiculous when I asked why they let men exclude them from the village watch. They answered that it was the duty of men. Egbisim men said they did not simply sit around on igot days but also performed duties they believed were beyond the capability of women. An often-cited incident was a house fire the igot men of Emomoro put out when the home's owner was away in the bush. As one said, "Go and set a fire in that house now if you see any woman there. They'll all run away with their hands on their heads." But a more pertinent reason men give is that they live close to enemy villages that could enter a village in the absence of men and burn it. So having men from age sets monitor the community daily was a precautionary measure.

This explanation, however, does not resolve the primary question of idleness, because some form of work could be done during the village watch. Since the only requirement for the igot watch is that the man be present at the lineage center for much of the day, a few men use igot watch time to weave mats for their homes or to carve hoe handles. But there is rarely more than one person at the watch center engaged in such activities although they are not against the rules. Many of the men I noticed weaving mats or carving hoe handles did not belong to the igot age set of the day; some had come to spend their work time in the company of other men at the center. Only two boat builders in Emomoro and two petty shop owners in Egbisim often engaged in part-time work activities during their igot. Boat building went on at one of the lineage centers and was always a part of the scenery of men lounging around, talking, or just sleeping on wooden slabs. The duties performed by igot men is important for maintaining peace and tranquillity in the villages, but the idle time could be directed toward some productive activities, because men seem to spend a significant amount of idle time just conversing, sleeping, or waiting for something to happen.

From my observation, women worked longer hours than men and their tasks were not lighter than men's. Most women walk an average of fourteen kilometers a day to and from their farms, they carry an average fifteen-pound load to the farm and return with an average thirty-pound load, they break firewood for cooking, prepare the evening meal, and do myriad other chores. Most men going to the bush carry nothing more than a machete, a gun, and a small bag slung over the shoulder. Except on a few occasions when they may need sticks for home repairs or ropes for weaving, they return with nothing more than what they took with them. After men's farmwork, men's loads are carried home by women and older children. In the fishing season, men dry most of their fish back at the fishing location, where itinerant traders visit them and buy up the fish. They carry home an average of a ten-pound load of fish primarily to be sold for immediate cash needs (such as the evening palm wine), occasionally saving a small quantity for the evening meal.

A more pertinent explanation for leisure time may be found in historical gender and age relations. From what we know about the Biase so far and from what we will find in subsequent chapters, such relations were characterized, among other things, by an abundant and available household and communal labor supply. Elders had little problem getting the cooperation of youths to do farmwork or to accompany them on trade journeys. This may have given elders ample rest from physical activities, maximized their role as cultural advisers and custodians, and reinforced their social status. Although this pattern was adopted by subsequent generations, it did not keep pace with the elements of wealth, fear, and respect for elderly men that originally fostered it. Consequently, although it was strategically important to preserve and perpetuate practices that reinforced respect for authority, women and younger men had become more assertive, traditional beliefs and ideologies were tested, scorned, and violated, and men's income diminished, along with their status. This had some social consequences for Biase villages.

It was no longer the prerogative of elders to sit around and ponder the affairs of their society; more than ever before young men dispensed with social decorum by engaging their elders in intense arguments at lineage centers. And the subsequent increase in the independence of youths meant an increase in the workload of women, since they could no longer rely on their teenaged and adolescent children (both sons and daughters) for domestic help. Women, prohibited by custom from sitting at lineage centers, assumed additional responsibility to care for the family while fathers, husbands, and sons were at leisure.

In recognition of our own complacency, Emomoro men met in May of 1990 and formed Okwu Onne (a meeting of love), an all-male association that meets once a month to discuss and be more involved in village affairs. Since youths blame elders for the economic and social stagnation of our villages, the older generation refrained from contesting for any of the offices of the new association, apparently to let the young prove themselves. The association set for itself the tasks of overseeing the cleanliness of the village, constructing temporary bridges over difficult village and farm routes, and providing for the general interest of the town. They predicted that the organization would be as strong as the women's meeting, although it had already existed for two months before the ferry point incident reported earlier. To signal its seriousness of purpose, they collected about three hundred Naira in fines from men who had been either late or absent during the last meeting; latecomers were fined fifty kobo, and men who were absent were fined five Naira. Although many offenders gave good reasons for their infractions, no reason was accepted as justification for people being absent from or late to meetings. A roll call is conducted at the end of the meeting and those not there to respond are considered absent even if they were present at the beginning of the session. These men are fined fifty kobo at the next meeting. Some people who learned that they would have to pay a fine for being late walked out of the meeting in anger, and others noisily denounced the fines and promised not to pay.

This trend continued every time I attended the meetings. Members of the younger age sets constituted the majority of those who either stayed away or were late for the meetings. Some who were absent from the current and previous meetings sent in their fines for the last violation and were expected to repeat the process at the next meeting. One of them gave this reason for his absence: "When I go there I just sit there and waste my time, because they don't have anything to talk about. Everyone wants to say something, everyone thinks their talk is the most important. If I go there and open my mouth, they all jump on top of me because they say nothing good ever comes out of my mouth. Instead of getting into a fight, I go fishing, make a little money, and pay them their fine." Such lack of interest in the young association should be expected because there has rarely been a strong organization to complement the functions of the Abu association; furthermore, the secular nature of the new association removes it from the sometimes-feared realm of the supernatural, which had been an important controlling element of male associations. It will be interesting, however, to find out how effective this men's answer to the women's association turns out to be, and what effect its operations will have on the power of the women's associations.

Effect of Gender Relations on Work Organization

In rural communities such as those of the Biase, it is presumed that care for children by members of the extended family is always available and releases mothers to attend to the many chores they face daily. While city women are likely to enjoy such luxuries because of the mutual benefits when relatives taste city life as well as provide help, rural women are less likely to have the help of relatives. Back in our villages, such relatives are required to be on the farms helping other family members or doing their own farm work.

Rural women constantly negotiate time between farmwork and child care, and because of the expectation that they will feed the family from their farm crops, are often forced to decide in favor of food production.[4] It is possible that this condition of neglect may have a negative effect on building the future generation of workers. Many children grow up mostly on their own, with little supervision. After they have been given their morning meals, their mothers leave them to play with their

4. For example, the pastoral Pibor Murle of Southern Sudan define an ideal wife as a woman who feeds her husband by being an industrious farmer (Andretta 1989).

friends and the mothers go off to the bush, returning only late in the evening. Throughout this period, some children are usually on their own and are not fed. Some mothers I spoke with said they always leave food at home for their children in the hope that siblings or friends will help them to eat. Many of the children are too small to feed themselves; besides, the type of food left is usually food the children cannot manage. Very often it is gari. People add water and some salt to it and then eat it by scooping it with a spoon or the fingers. Since most of the children are between three and four years old, they find it difficult to prepare the gari and some of them go without food until their mothers return in the evening. Sometimes elder siblings between five and six years old are left to take care of their younger siblings. The children tend to grow up not expecting more than two meals a day, although they always clamor for food when they see it.

The eighteen-year-old mother of Etan often left a small portion of food for her son before going to the bush in the morning. Neighbors believed she was training him to be independent while she was away on the farm, because there was nobody to take care of him. Etan would go around with other children whose mothers also left some food, joining in the meals without waiting to be invited. I observed him on three randomly chosen days to find out how he was responding to the "training." He did not play much but seemed to cry less than most of the other children. Whenever any of the other children brought food, he made sure he was sitting right in front of the bowl and would start eating although there was never any formal invitation to eat. Then he would wander into his mother's house and drink from the water pot. When he felt like defecating, he would simply squat wherever the need arose, relieve himself, and wander off again.

Etan's case is neither unique nor restricted to the homes of single parents. Rural men and women leave home early to start their daily routines of farming, fishing, gathering, talking with friends, and attending the neighborhood markets. In many cases there is nobody left at home to take care of the children. Some households have elderly persons who care for children but this is infrequent, since the old people themselves must attend to their own livelihood. But any elderly person around is likely to be watching children from up to five households, including children hurriedly left by mothers who disappear through the back door before their children discover they have left. Sometimes a mother leaving her child will simply shout, "Mama please listen out for Ogbor. I'll only stop at the cassava farm to get some food and will be back shortly."

Some mothers take their children to the farm if they anticipate light labor such as picking pepper, but weeding or tilling the ground is difficult with a child tied to the mother's back. A few mothers take younger siblings along to the farm so they can keep an eye on infants while the mothers work. One mother said she had discontinued the practice after her daughter ran away from a snake she thought was chasing her, leaving her infant sibling a few feet from the snake.

Mothers begin to prepare food when they get home and the evening meal is ready to eat about an hour later; most men are back from the bush and have gone to find some fresh palm wine or to eat in the homes of concubines.

Low food consumption is not limited only to children but appears to be patterned into the Biase livelihood. Almost all farmers take some food to the farm or cook a meal there, but this meal, eaten when the farmer decides to take a rest from work, is usually his first meal of the day; he eats his next meal at home in the evening.

Although the Agwagune seem to have little problem with the availability of food, 90 percent of the households in my survey reported eating less than three meals a day. The most common reason people gave was that they were in the bush most of the day and had little time to think about food. In one of our informal talk sessions, men were unanimous in blaming women for this predicament and the trend of their opinion was predictably androcentric: women no longer respect their husbands the way their mothers used to; they are dissatisfied with their husbands' lack of effort and are driven to affairs with other men, which is why many of them are without husbands; women eat three meals a day by taking food to the farm, but none of them prepares three meals a day for her husband. Two women breaking melon seeds nearby interrupted and retorted that women spend most of the day on the farm and cannot cook before they come home with food from the farm. They said that most days men only go and check their fishing nets in the morning and then come back and sit at the lineage centers all day waiting for women to return from the farm to cook for them.

Couples occasionally quarreled over the feeding of their children. Neighbors rescued a woman, Eveghe, from her husband who kept hitting and screaming at his wife for returning late from the farm and leaving their three-year-old unfed since a late morning meal. He explained to people who had gathered in his house: "It doesn't matter to me if she decides to keep me hungry for a week because if I come back and find nothing in the pot, I will go to the house of Okpa and look for two cups of palm wine. After that I go to sleep most days without eating anything at night. But what about her child? Since I came back, this child has been following me everywhere because she is hungry. Will I go and cook for her with my fingers?" Some women who were present advised the woman to plan to get home much earlier to avert a recurrence of the incident. They then pleaded with the man to be calm, and explained that it is sometimes difficult to know how late in the day it is when one is working on the farm. In a small group that gathered later away from the incident, neighbors said Eveghe's husband never brings any food or fish back from the bush for the family to eat. He sells his fish on the way home and heads directly to the palm-wine stall. One said Eveghe was a hard worker and that the family would die of hunger without her.

This incident points to the classification of male and female contributions to household sustenance, or to the status of food crop contributions versus monetary advances. The general pattern of food provisioning in households is geared toward what a woman brings home from the farm or garden. On a daily basis, a woman brings home cassava, obiara, and vegetables for preparing sauce. If she is unable to go to the farm to replenish her food stock, she asks a friend or a neighbor to lend her some pepper or vegetables, which she will repay promptly. It is her responsibility to make sure that her farms have sufficient food for daily consumption. If she is perceived by her husband as not fulfilling this duty, she is considered an unsuitable wife and mother. The Biase husband feels particularly justified in his anger if, as it is often the case, he "divides" his farm for his wife (that is, lets her intercrop corn, cassava, or obiara on his yam farm). Such gestures by men signify their care for women and are regarded as a major contribution toward the food procurement of the recipient, comparable to the provision of a large capital investment. In Eveghe's husband's perception, his wife is running the business poorly. This notion is fueled by his realization that he had spent money hiring labor and buying drinks for the friends who helped him prepare his farms for cropping.

My discussions with men and women showed that money is valued more highly than the contribution of a food crop. Since men were presumed to be provisioning the household's monetary needs, they logically saw their role in household sustenance as more important than the generally acknowledged greater contribution of food by women. If a man had an unsuccessful day at fishing and came back without fish for the evening meal, he felt a sense of accomplishment when he gave his wife money for fish and expected his wife to express her gratitude. Men always accuse women of hoarding money and of constantly demanding money from their men for small household needs. I recorded seven incidents in one month when couples fought over the provision of money. Two of the incidents resulted in separation: one woman moved her belongings back to her mother's house, vowing never to return to her husband; another took her case to the lineage center, where a group of men sat.

> All of you who are friends of Ogban, do you just go back to your concubines every night and ask them to bring food and open their legs for you? I am talking about one piece of fish, just one piece. When you see Ogban ask him if the pepper and salt and cassava he eats in this house every day have come from his pocket once in the past month. Ask him if he knows where the soap he bathes with comes from. I am telling all of you this now because if he comes to my house again, I don't want anybody to say an ordinary woman broke Ogban's head. All of you who are his friends are hearing me now. Go and tell him to sleep wherever he is.

Such outbursts are common and help redirect the attention of husbands to their responsibility; the resulting separations are often temporary. The men pleaded with the woman not to be angry and to settle the matter peacefully with her husband.

These cases, however, confirm what has been recorded for other cultures, such as the Mossi of Burkina Faso, where "the income from a woman's cash crop production is used to satisfy requirements of school fees, clothing, and medical supplies, as well as to pay for additional condiments for meals" (Gladwin and McMillan 1989: 351). Women's contribution to domestic sustenance is therefore not limited only to food crop provision; often it extends to providing money for supplementary household needs that are neglected or ignored by husbands.

Conclusion

The consequences of the diminished authority of Biase mystical associations extend beyond the effacing of male authority and the emergence of female power. Partly because the men's associations were pivotal in social cohesion, particularly kinship affiliations, their demise left little structure for consolidating kinship obligations. Other factors no doubt account for the declining importance of kinship obligations in Biase. But Schneider's report at the beginning of this chapter that a group of brothers refused to assist their old mother in her farmwork is indicative of the threat to the cooperative ethic among some family members. In the case of the Biase, the lack of a coalescing center may be attributed to the resentment community members felt toward elderly men and a disaffection with the operation of the mystical associations. The feeling expressed by some villagers that their relatives in the mystical associations were responsible for certain family calamities fueled the local belief that Inyono members sacrifice the interests of kinship to maximize the selfish economic and social benefits offered by mystical associations. The new challenges to traditionally sacred institutions combined with lessening of wealth to significantly lower the men's authority to enforce social rules.

This new social configuration seemed inevitable because of the men's dogmatic adherence to the supremacy of their Abu association. This allegiance prevented the emergence of secular associations for men as alternatives to the declining influence and authority of Abu. Consequently when the authority of the mystical associations declined, there was no framework for men to assert their roles that were sanctioned by custom. On the other hand, women recognized the limiting nature of Egup and sought early to curtail its powers by forming a secular association that recognized the importance of everyone in its management and sustenance. This background gave the Aka-e-Mitin Aneba a strong basis for women to cooperate more effectively than men.

SEVEN

Rural Politics in a State Polity

> The police drove into the village in one huge truck. One of them came up to me and asked, "Are you one of those we are looking for?" I said I didn't know what he was talking about. Before I finished the sentence, he landed a heavy slap against my face; he kicked me and said I was pretending to be blind. Then he pushed me down in the mud. Then I heard Onun Obazi Esu scream. It turned out that one tiny policeman about the age of the Onun's grandson had conked him on the head. The boy kept shouting to Onun, "Shut up! shut up!" Then I heard another slap. My heart was boiling. Can you imagine the police slapping the Obong of Calabar? These children came here and slapped our Onun. If it weren't for my ailing leg, I would have done something bad that day.
>
> (Uka "Spider" Ejituru, 46, an Emomoro informant)

This incident occurred in 1986, four years before I started my research. It underlines my earlier suggestion that because of their isolation, poverty, and powerlessness, groups such as the Biase are held in low esteem by the state. They seem reluctant to interact with state representatives for fear of violating decorum in ways which might result in the type of experience cited here.

I have noted that Biase political organization revolved around the authority of elderly men respected and feared because their role as patrilineage and village Inun had supernatural sanction. Although this political organization served the Biase well for years, it currently faces significant stress under a nation-state. The Biase face the ambiguity of operating traditional laws under the control of statutory laws, but are so deprived of state-provided amenities that access to government services is virtually absent. This change resulted partly from a restructuring of authority relations within once-powerful communities and interaction with external influences, and partly from the effects of colonialism, which directly affected the Agwagune, who were attacked by a British expeditionary force contesting control of coastal Biase communities. The Biase believe that this changed political structure is a direct cause of their inability to improve their situation. In this chapter I will first examine the changes that have occurred in Biase internal and external political arrangements and then assess their impact on the use of resources and the interface between state and local political traditions.

The Effect of Incorporation into a Nation-State

At the time of my research, the Biase belonged in the Cross River State (fig. 1), one of many states carved out of the political and geographic block that used to be known as Eastern Nigeria. That year its political affairs were administered by the Akamkpa Local Government, named after the small rural village, Akamkpa, in which its administrative offices were located. The location of Akamkpa (40 miles from the state capital, Calabar) was, however, closer to the state capital and more easily accessible to state agents than to Biase villagers. The Biase Local Government Area (LGA) (fig. 2), with offices in Akpet, was created soon after my research. This local arm of the state government is administered by a council of elected representatives of the Biase. It is too early to know what effect the creation of the Biase LGA will have on the people particularly because the uncertain economy of Nigeria has forestalled any development ventures. The immediate result of creating the new LGA is the generation of a few wage-labor jobs for Biase youths seeking alternatives to farmwork; their labor mainly involves clearing village paths and constructing temporary wooden bridges. The euphoria over the creation of a separate LGA for the Biase and the novelty of electing and being represented by their own people have so far muffled any criticism of the elected representatives. Besides, it is unlikely that at this point most Biase know exactly what to expect of their representatives or that their representatives know what to expect of the state.

The inclusion of "simpler" cultures in the machinery of the state results in rudimentary indigenous rules operating within the precepts of complex state requirements. Among the Biase, this mix between traditional and modern results in the transference of authority from traditional loci to state institutions and tends to set up powerful stresses in the management and regulatory systems of the traditional society. This condition is succinctly illustrated by Biase villages, such as Abini, that are located along state highways.

Biase men sling their guns and machetes over their shoulders when they leave home for work in the morning and come home in the evening. Until a few years ago, this practice never conflicted with the state requirement for possession of a firearms license because it is primarily confined within the limits of rural boundaries. With the completion of the Calabar/Ikom highway this boundary broke down for communities located on state highways. Abini men crossing the highway to their farms, with their machetes and guns slung over their shoulders, became exposed to state patrols. The increasing contact with travelers from other communities and state agents prompted the state to bring Abini, like other rural people in similar circumstances, under Nigerian state laws that regulate the use of firearms, including flintlock guns. This move resulted in a reassessment of hitherto unquestionable rights and privileges enjoyed by the villagers as their traditional laws were superseded by state laws. It was no longer legal for men to simply pick up their guns and go; they now had to invent strategies for eluding the police without giving up an important tool of their sustenance. Control by the state tended to further weaken local political authority (table 12) and the Inun appeared to be incapable of protecting traditional rights. This change in the political system of the Biase was accompanied by diminished respect for traditional cultural standards. Power and authority relations based on gender and age, as well as on reverence for social norms, were no longer accepted; at the same time regulatory associations such as Egup, Ekpe, Ebrambi, and Abu lost their preeminence as institutions of social control. Historically the power of such regulatory associations, especially Ekpe and Abu, was strengthened by an overarching membership structure that extended across neighboring groups (Ubi 1986; Attoe 1990); a member of the Ekpe association in Calabar was equally at home at the observances of the Ekpe events in Agwagune or Ekoi. Thus culturally unrelated southeastern Nigerian groups, such as the Efik and the Ekoi, shared allegiance to the codes and symbols of the Ekpe cult that conferred social privileges on members irrespective of their geographic regions. This inclusive membership structure made it possible for Ekpe to serve as a political and economic broker between villages in the Cross River area: "Ekpe ensured community orderliness and smooth commercial transactions between traders of neighboring communities and the Efik. Laws for the orderliness of the communities were formulated by Ekpe, and members and nonmembers of the cult obeyed. Cases were brought before the cult and satisfactorily settled, and the head of Ekpe became, in fact, and in law, the head of government" (Ubi 1986). The assignment of power and authority to local associations significantly weakened the role of the political Onun, par-

TABLE 12 Residents' Assessment of the Inun's Authority in Odumugom

	High (%)	Medium (%)	Low (%)
Success in enforcing social rules	19	23	58
Respect within Community	30	40	30
Respect outside Community	53	20	27

Note: $N = 30$

ticularly in Agwagune where the political Onun and the Onun Abu belong to different patrilineages. The position of the Onun as the political leader of the community was culturally separate from that of Onun Abu, although both officers exercised judicial functions within the same community. While the community Onun is nominated by the appropriate kingship patrilineage and appointed by the consensus of all the village Inun, the Onun Abu is assigned to his position by the patrilineages traditionally authorized to do so.

The conflict implicit in this arrangement was marked by the eclipsing of allegiance to the community under allegiance to Abu. The political Onun, by virtue of his cultural position, was not only a member of Abu but belonged in the upper Abu hierarchy. This often resulted in communal interest being overridden by Abu interest. When faced with a situation in which one interest competed with the other, Abu often won the confrontation because the authority of the Onun also derived from his membership in Abu or Ekpe. I explained that members of Abu refused to have the community mediate an ongoing dispute among their members because they felt the town should not interfere in the affairs of Abu. This exclusionary privilege of Abu often extended to individuals, who used the Abu palm-frond symbol to protect their property from use or to bar fellow community members from carrying out a collective decision. One day a dispute arose over the ownership of a fishpond. After the villages had fixed a date on which to fish the pond, an Emomoro man claimed that the pond belonged to him. To keep anyone from going to the pond, the man placed an Abu palm-frond across its entrance. This claim was strongly contested by women in the area, who took the case before the village Onun. Since this matter concerned an individual Abu member rather than the association, the Onun council settled the matter quickly and asked that the Abu symbol be removed. When Abu later lost its authority base, the effect inevitably spread to other levels of traditional polity. Most notable were the challenges to ruling patrilineages over the appointment of community Inun, the diminished authority of elderly men, and the ascendancy of younger men and women to authority positions.

The Effect of the Local Polity on Social Processes

In Agwagune the challenge to the prerogative enjoyed by the kingship lineages was marked by a redefinition of eligibility criteria by members of nonruling houses. Traditionally the only relevant criterion is that the Onun be the eldest man in the lineage. This stipulation on previous occasions took precedence over other qualities, such as the moral character of a nominee. It was expected that the importance of the new role would reform individuals of questionable character and steer them away from their previous unethical behavior. Resentment over this inflexible rule became apparent after the death of the last Onun Agwagune. Onun Odidi Okpitu was generally seen as embodying the characteristics of our past Inun and as capable of restoring our former glory. He had a flair for getting things done with little objection; adventurous neighbors were promptly subdued. Since his death in 1988, no new Onun has been installed because nonruling lineages have persistently challenged the competence of nominees. Some believe there is no one left in the kingship lineagės deserving to be Onun. Others publicly advocate that Agwagune should not rely on the kingship lineages for the next Onun.

To contend with the absence of a political head, our village and patrilineage Inun formed the Inun Council (plate 25), a temporary caucus of Inun that manages affairs pending the appointment of a new Onun. The council directed early in 1990 that the kingship lineage should present an Onun-elect at its next meeting; this did not happen while I was in the field because the person favored by the lineage was not favored by most members of the council.

Outside the kingship houses, the challenge to tradition mounts, since many want the next Onun to come from any lineage irrespective of kingship connection. They contend that the inflexible custom of having Onun only from kingship lineages has resulted in the political weakness of the Agwagune, our loss of prestige among our neighbors, neglect by the state, and the lack of economic growth. At one of the council's meetings, an advocate of political change said,

> People who have little tact in manipulating the political situation in the country are presented as Onun. Nigeria doesn't

> want people like that today. Unless we change we will continue to remain in this mud with no one ever knowing about us. You bring someone who does not even know the way to Calabar and say he is my Onun. There is no direction here. These people don't know how we can become like the Igbo. How we can talk to government about our economic slump and get money.

The thrust of discontent seemed purely economic, particularly because of the harsh times all Nigerians currently experience. Not surprisingly, a prominent business man from Emomoro, popular for his generosity, led the call to have the Onun come from non-Onun lineages. He was favored by many in Agwagune because they believed his extensive contacts within and outside the community make him a more suitable candidate to help our people to improve their situation.

Opposition is strong. Kingship lineages and community elders concerned with preserving the system strongly criticize such advocacy of change. People from the ruling lineages have sworn to resist any attempt to deprive them of their traditional right, hinting that anyone who fraudulently became Onun would die. Traditionally when an Onun lineage presents any of its members to the community as a new Onun, all our patrilineage and village Inun meet to formalize the appointment in a confirmation ritual; there was often no protest from community members or from any of the other lineages.

Abini was experiencing a different kind of Onun tussle featuring personalities from a kingship lineage and a nonkingship lineage. The dispute was fueled by a new government policy that was soft on appointment of traditional heads. The state encouraged communities to nominate members who would serve as community heads and be formally recognized by the government. This was the government's attempt to minimize the disparity that exists between the leadership structures of many Nigerian communities and to bring previously neglected groups closer to government's attention. In compliance, some rural communities designated lineages as village units, suggested the names of patrilineage Inun for appointment as new village Inun, and got easy government approval to appoint a multiplicity of Inun. There were frequent accusations by Biase villagers that some of the new Inun who were never nominated by their communities paid government officials to issue them certificates.

A series of conflicts between lineages and individuals has resulted from this arrangement, disrupting cooperation between community members and the new Inun. In one case, a non-government-certified Onun, who is also the more popular one among the people, convinced the government to provide the town with boreholes. When work started, other village people sympathetic to the certified chief arrested their fellow villagers who were helping out at the project site. They accused workers of illegally digging up the town. This further escalated tensions within the villages, because some of those arrested were temporarily locked up in improvised jails until family members paid the fines imposed by the certified chief. In another case, the construction of a village bus terminal was halted because the project had been initiated by the uncertified Onun. This interrupted project became a monumental testimony to the animosity within the villages and a constant reminder of a failed attempt by locals to improve their economic circumstances. Work materials bought with money from local contributions were strewn all over the ground, which at the time was covered with thick brush. Informants said they had been left there since about a year before my arrival in Biase. There was every indication that the project would not continue until the power struggle in the community was resolved.

A government-appointed chief blamed the conflicts within some Biase villages on uncertified Inun who sway popular opinion against certified chiefs. Uncertified Inun felt pressed to outperform their certified counterparts in order to justify and maintain their popular support. The certified chiefs, meanwhile, are unable to build much local support and have little influence and sometimes no access to the government officials they represent. Consequently community projects, such as clearing the farm routes, are more often successful when initiated by those Inun who have overwhelming support from their members.

Such conflicts distract Biase villages from maintaining some of their services. For instance, the nine-kilometer route linking Odumugom-Agwagune and Adim, along with the trail to Abini, is usually cleared by community people at the beginning of the dry season, in anticipation of increased travel. When I arrived in Biase in January, only a few meters of the routes had been cleared. Most were covered with brush, making traveling difficult. Traveling along village routes overgrown with brush is both inconvenient and unsafe. In the dry season territorial encroachment is often at its peak as neighbors wonder freely across each other's borders. The incidence of hostility is higher during this period, as villagers encroach upon each other's territory to fish the ponds, tap wine, exploit forest resources, and farm in the bush, than it is in the rainy season, when flooding limits most activities; as a result, villagers tend to be more protective of their territories and more watchful for danger in the dry season than at other times. There were recently two fights between Agwagune and Adim and one between Abini and Idomi in the dry season when the respective villages went to mark boundaries in preparation for the beginning of farming. On a few occasions (the most recent being 1986), hostile neighbors hid in overgrown brush and ambushed and killed travelers passing through an enemy village. The Agwagune often refer to the case of a man who was ambushed by the Adim on his way back from his farm. After a second man he was traveling with briefly stepped into the forest to tap his palm-wine, some men suddenly emerged from the bushes and dragged the victim away. When his companion came out a few minutes later, he found a hand severed at the wrist lying on a heap of sand beside the farm route. The severed hand turned out to be his companion's. This incident later culminated in revenge attacks against Adim and discouraged people from using the route even after the brush was cleared. Routes are safer when people farm by the roadside because much of the brush is cleared over a wide area, making it possible to detect unfriendly action sooner. Where people pay little attention to such clearings, farmers, fishermen, and other users of the forest are frequently ambushed.

Elders and youths blame each other for the neglect of the community, and they often exchange strong words in local meetings aimed at solving the problem. Younger age sets say their elders are ineffective and lack the skills needed to manage the community under present realities; elders say the young shun their wisdom, tending to look more toward city attractions than traditional ways. There were numerous fierce quarrels between one age set and another over their respective competence in managing communal matters. In one case, a member of the Guinea age set (ages 40–45) accused a member of the Africa set (65–70) of being responsible for all the failed projects in the village. He cited a school-building project that was abandoned by the Africa age set halfway through the construction as evidence that elders were unresponsive to the needs of the village: "If you say you are the elders of this village, tell me why that building has remained the way it is for the past five years. Your only concern is to wake up in the morning and go and check your fishing. Even if this village were on fire, the first thing you would do is put your gun on your shoulder and head into the forest. All of you now live in the forest. When strangers come to this village they always think that all our old men have died." Traditional formalities would dictate less display of such youthful outrage, but recently decorum has been routinely abandoned by youths in their dealings with village elders. The above accusation generated a caustic reply from the outraged elder who, like many of his age, seems continuously befuddled by the effrontery of youth:

> If the hair in your nostrils does not turn gray and dangle around your ankles you'll never know that you are also responsible for this village. Why should an old man like me continue to labor for a young man whose feet are stronger than mine? When I started contributing my labor in this village your mother was not even born. Go and ask your uncle if I was not one of the first people to cut the first trees that were used to construct the bridge where all of you now walk with pride and make all your money. Without my work, you would not be able to stand in front of me today and insult me. Go and ask your uncle, go and ask him; ask anybody in my age set.

Conflict in local polity is not only manifested in inter–age set confrontations but is also endemic in other social groups. Once the Onun of Emomoro village invited me to a meeting of the Inun of his patrilineages which had been convened to discuss the visit of a tax collector. When I asked the Onun why only three of the seven Inun were present, he explained that he had no one to send to summon the rest of his lineage Inun. One of the Inun present quickly interrupted him: "What do you mean by saying there is no one around you can send on errands? This is the first time I've heard an Onun of a whole Emomoro village say he cannot call on anyone in his village and ask the person to summon Inun to his house. If you don't have such authority why did we make you our Onun?" The Onun Emomoro turned to him and said, "All right then, I pick you. Go and call the others and tell them we have been waiting." There was an immediate objection: "How can you pick me? Did I do something evil when I said that you could have called on anyone? There are many children running around that you could ask. Why do you pick an Onun like me to run such an errand?" The offended Onun left the room, expressing his displeasure as he walked away. The meeting was later canceled because no other Inun came. When the tax man came the following day, the Inun had not met to decide on how to welcome him; the food, drinks, and money usually offered to mitigate the harsh tax assessment were not ready. When the tax man left, the elders and teenagers who had been exempted from the tax the previous year were now enrolled to be taxed. The younger age sets expressed their indignation by refusing to participate in communal labor, saying they needed the time to work for tax money. For a while there was resentment in the village. More than ever, the youth believed their elders had failed them.

The Emergence of Alternative Political Groups

Under such circumstances, local organizations such as the women's association, secular associations, age sets, and a caucus of Biase elite living in the city assumed a number of responsibilities that had customarily been performed by the head onun and his council. People transferred judiciary matters traditionally handled in the Onun's court to alternative social groups because of the waning authority of the elderly men in enforcing decisions. Biase women, for example, have become increasingly dependent on their association to mediate judicial matters. When a dispute involves only women, the association can easily resolve the matter because compliance with association rules by members is very high. There was no case of anyone refusing to pay a fine or carry out a restitution directed by the Aka-e-mitin Aneba. When women are unable to pay fines imposed by the association, they urgently ask friends or family members for the money.

On the other hand, people regularly challenge the directives of the Inun council. For instance, the Odumugom Inun council decided a case of theft in which the accused were asked to pay six hundred Naira each as restitution for stealing mattresses from an unused community health center. One of the community elite, a close relative of one of the accused, challenged the decision and threatened to take the Inun before the state court for not letting the courts decide on innocence or guilt. The Inun withdrew their decision, took no further action, and apologized.

On other occasions, people have transferred cases pending before the Inun council to the men's or women's associations. Although men's associations have lost significant power since the days in which they performed major judicial functions, the vestiges of authority that justify their existence are still perceived as more effective than what litigants get at the Inun council. Abu dispenses with the lengthy hearings and arguments that characterize the more democratic Inun council. The presence of any two key Abu members is enough to initiate an Abu hearing; non-Abu members are barred from the proceedings—women litigants and their supporters remain outside and speak only when requested to; interruptions and arguments are censured; and the strict authority hierarchy within Abu permits quick judgment with little dissension. Consequently Abu is reputed to resolve conflicts and enforce compliance with judgments faster than the Inun council.

Despite the seeming authority of Abu, informants said that the Abu of some villages is less effective than others; the strict codes by which the association operates have been weakened by filial sentiments. In some villages, Abu members dispute decisions that affect family members. Litigants have transferred their cases from the Abu of one village to that of another with a better reputation. On a few occasions, villagers have attempted to ensure favorable judgments by threatening to transfer their cases. As a result of this weakening political influence, some indigenous associations are also losing power to more conventional ones. Young community men have started secular associations to cater to their interests as a response to the diminished influence of men's associations.[1]

Although the young attribute the neglect of the community to their elders, it seems that authority even among the younger age sets reflects the weakness at the communal level. Congo and Guinea age sets, for instance, often conflict with younger age sets (Unity, Freedom, Tunisia, and the adolescent groups) when they supervise them at communal work sites. They complain that the older groups do not censure people who were absent from previous communal labor. In one lineage meeting, Emomoro adolescents told the older age sets that they would no longer respond to calls to do communal work because their peers who had stayed away from previous work were not punished. On another occasion, members of the Unity age set refused to clear the brush around the primary school because some of their members who were

1. The traditional Ebrambi warrior group faced such a dilemma. Historically, Ebrambi "was the military arm of the government in parts of Biase land such as Umon, Agwagune and Adim. The basic function of Ebrambi was to defend the boundaries of a unit's territory and also to fight wars" (Attoe 1990: 59). In recent wars, however, Ebrambi has played only a minor role serving either to instigate or prepare warriors to fight through song and dance or to celebrate victory. Its current function is primarily ceremonial since Ebrambi is a major feature in communal festivities and observances. However, a community crisis developed in Okurike when their young men founded an alternative group to the Ebrambi. After a war between the Agwagune and Adim in 1986, Okurike men who participated in the war banded into a warrior group intended to be an active military arm of the village. The advice of angry community elders that the group should disband was ignored. When every member of the new group mysteriously became ill, the rumor was that community elders with strong affiliations in Biase mystical associations were trying to scare them into submission. The lack of popular support from other Agwagune youths, however, prevented the group from gaining the momentum its founders hoped for.

absent from a previous labor call were not disciplined by the work supervisors. The younger age sets accused Congo and Guinea of being too weak to organize labor, enforce compliance, and punish offenders—the same complaint Congo and Guinea make about the older age sets.

Communal Development Programs

In spite of these conflicts, Biase age sets have contributed significantly to the economic growth of the group. Abini age sets have built market stalls for the town; the Guinea age set of Odumugom-Agwagune built the local customary court for the Egup-Ipa subgroup of Biase; and the town hall is being built through the collaboration of age sets and other social groups, including the Aka-e-Mitin Aneba. The problem with projects such as these is that the primary motivation for starting them is communal prestige rather than filling an important communal need. Agwagune informants noted that when Agwagune was selected by the Egup-Ipa subgroup as the site of the customary court, they intended to honor the Onun Agwagune at the time, Onun Odidi Okpitu, because of his prestige among the Biase. The Agwagune have not been so honored because, as was noted earlier, diversion of trade and transportation from the coast has led to the loss of their prestige, and the internal discontent of younger members has crippled the power of elderly men.

Biase communities tend to favor projects that have minimal growth impact but symbolically elevate the beneficiary community's status among its neighbors. According to customary court officials in Odumugom villages, the Egup-Ipa court building has brought no revenue to Odumugom villages since its construction over twenty years ago. The additional local income that would have come from business of users of the court has not been realized. The court was built about three kilometers from the village, beyond some of the more dangerously flooded segments of the Adim/Agwagune route. The purpose was to minimize traveling inconvenience for members of other Egup-Ipa villages attending court sessions during the rainy reason. Ironically, the inconvenience was transferred to the Odumugom villagers who were separated from the courthouse by the flood plain; some of them have missed court sessions either because they could not walk so far or because the route was flooded. Contrary to our initial expectations, local food vendors got no business because many users of the court brought their own food when they attended court hearings rather than walk the long distance to the village for meals or brave the seasonal floods. An additional burden on the villages is the frequent need to clear brush at the site. The Onun of Emomoro village said that if the area is left unattended, other Egup-Ipa communities may suggest that the court be transferred elsewhere. But the village gets no rent from the government for having the subgroup's court in its area.

Another project was the Agwagune town hall which had been under construction for twelve years. During my research, the abandoned one-third of the construction was invisible from the road because it was buried deep in the bush. Informants said work was discontinued because of the high cost of building materials at the time; all the work that was accomplished was by the effort of the villagers who contributed their time, labor, and money toward completing the town hall. The reason for the location of the town hall away from the village was originally to fit it into the plans for a projected new site of Emomoro/Egbisim villages. Because of the problem with floods, the two villages decided to relocate north of the flood plain where the problem was less devastating; but only a few families have moved to the new site. There seems to be little encouragement for those not living at the new site to spend more time and money constructing a facility that is not likely to be used for a long time. Consequently people have been unable to contribute further labor, time, and money to the project.

The Abini also make little use of the market stalls constructed by their age sets. Many vendors prefer to spread out their items by the roadside where they can easily be seen by passing motorists who stop and buy them. Some stalls are not occupied on market days, and many others are underutilized. Although a member of the age set that constructed the Abini market stalls said the stalls were intended as a major step toward bringing trade to Abini, it is apparent that the underlying motivation was group prestige: "Even if we die today, we have done something for this community. Nobody gave us one penny. All the money came out of our own pockets. If our women want to waste it, that's their business. But I can tell you that all Abini are happy about what we have done. The only problem is that we have no money to do big business here. We do not use things like canned fish, tea, bread, and toothbrushes. How many people in the village clean their teeth with brushes?" The major constraints the Abini face in maximizing the use of their market facilities have already been discussed: the absence of men from the market economy, reliance on maximizing social relations rather than economic benefits, and the burden placed on women to sustain the market economy.

The Response of Biase City Elites to Local Needs

While Biase villages tend to take up projects aimed at enhancing communal prestige, they depend on their elites in the city for initiation and implementation of projects that would generate economic growth. Onun Effime Agbam of Igbadara lineage confronted me one day with the following remarks: "People like you who see the face of government are not doing anything to help us. In this community we have doctors, lawyers, judges, university people, and even people like you who have seen how people in the spirit world live.[2] Yet you are not bringing us any of the things that you see other people do. You are all content to come home and defecate in the bushes, wade through mud and flood, and remain in this darkness like the rest of us."

In spite of these criticisms, various associations of the Biase elite in the state capital of Calabar regularly lobby government agencies on behalf of their communities. Ethnic lobbying groups have long been important in the involvement of Nigeria's village communities in the state political process (Diamond 1988). This is more true for minority ethnic groups that are politically and economically powerless. The Agwagune caucus in Calabar, headed by a university professor, Solomon Ogbodim Unoh, has been unrelenting in making representations to the government. For many years Unoh and his caucus unsuccessfully lobbied the state government to provide borehole drinking water, improve the health care delivery facilities, and approve the opening of a community-built postprimary institution. Such failures lend weight to criticisms of villagers unable to understand why people highly venerated back home do not attain equal esteem among important government officials.

Sometimes the need to follow government protocol in presenting a community's problems to the state has led to conflict between city caucuses and the villagers. Usually a caucus would write directly to the government and describe the problems of the community. If the signatures of Inun back home were required, the caucus would send a draft of the document to them for approval and signatures before presenting the petition to the government. During my research, the Agwagune caucus in Calabar believed it was important to get Inun directly involved in a report it was making to government concerning the encroachment of a

2. Reference to the United States and Europe.

neighboring community on Agwagune land. The dilemma the caucus faced was that the letter required the signature of an overall Onun of Agwagune if it was to get the attention of the relevant government department. But Agwagune was still without an overall Onun. The caucus believed it had solved the problem by appending at the end of the letter the names of nine lineage Inun and designating one of them as Onun Agwagune. When a courier brought the letter home for approval, some Inun interpreted the designation of an Onun as meaning that the caucus had decided who should be the next Onun of Agwagune and had imposed a choice on them. They refused to sign the letter and sent it back to Calabar. When he learned of the misunderstanding, Unoh hurried back to the villages and clarified the situation to the Inun. The letter was then amended and signed.

Onun Effime's remark quoted above, that a few Biase people occupy relatively distinguished positions, is correct, particularly when these distinctions are compared to rural standards. Rural people respect city dwellers more than villagers. Wage employment is almost synonymous with higher income and prestige. People who work in government offices are assumed to be powerful enough to significantly change the political and economic fortunes of their villages. The few Biase people occupying middle-level government positions are not of much consequence in a polity where, according to Berry, access to the state is "a precondition for doing business successfully" (Berry 1989: 44) because, "in competition for access to both local resources and state-controlled resources, the returns to an individual's investment in social relations depend on the wealth and/or influence of the group or its leaders" (1989: 48). The minority status of the Biase in the Cross River State polity tends to exclude them from appointment to influential civil service or political positions. Informants said some Biase who at various times had occupied high government positions had served in departments where they could do little for their people. Some were said to be reluctant to help get their people jobs for fear of losing their own, especially under successive military regimes that followed little protocol in terminating appointments.

Conflict between Government Projects and Local Needs

Just as the rural Biase get more involved in projects that enhance their prestige than in those that improve their circumstances, so when the state has intervened in rural development it has focused on projects that ease state operations rather than on those that promote communal growth. Government projects in Biase have had low budgets and often were of more importance to the government than to the people.

For instance, the Agwagune and Adim police stations were built to check hostilities arising from the boundary dispute between Agwagune and the neighboring Adim town which persistently endangered the lives of travelers. Both police stations operate from private residences and are rarely used by the people. Villagers prefer to take cases to local associations. The police often complain to the village Inun that their presence in the town is redundant because no complaints are ever taken to them.[3] Policemen sometimes patrol villages listening for fights or other disturbances that may be construed as criminal even if no formal report is made by villagers. For example, when the mattresses were stolen from a local health center, the nurse in charge immediately reported the incident to the village Inun, who started investigating the matter. The police later threatened to arrest the nurse if she did not make a formal complaint about the loss so that possible suspects could be arrested. The nurse refused to comply with their request, saying it would be disrespectful to the village. The police went ahead and arrested the people suspected of buying the stolen items although they had received no formal complaint. Informants said they rarely take complaints to the police because some villagers have had bad experiences with policemen. A commonly reported incident involved a group of policemen who stopped a traveler along the nine-kilometer route and robbed him of ninety Naira. It was also reported that policemen always demanded money from complainants before taking action on reported cases. Some said they had been detained over minor complaints and made to pay large sums of money before being released.

It was surprising, however, that the patronage of some of the health centers was also low. The town crier often went around urging villagers to take their children to the center. Biase villagers say they make minimal use of the few health centers in their villages because the services are inconvenient and expensive. Nurses often gave them prescriptions to buy drugs twelve miles outside of the village because the centers were always out of drugs. When I arrived in Biase, the Egbisim center had been closed for about two months; the nurse in charge had been transferred, and her replacement was away being trained. Health care had also become increasingly expensive. For an ailment like malaria, the pa-

3. Between January and June, the police recorded only four cases.

tient could get herbs from the local medicine man for about 80 percent less than it cost to be treated by the nurse, who often charged patients a twenty-nine-Naira registration fee before treatment. Rather than put much hope in such services, we preferred to treat ourselves or to visit local medicine men whose medications were inexpensive, readily available, and usually effective.

The government also embarked on an ambitious project aimed at easing transportation difficulties in the state. The government built bus stops on every six-kilometer stretch of the Calabar-Ikom highway for passengers traveling on government-owned buses. As in other cases where projects aimed at rural peoples were unrelated to their needs (Painter 1987; Nyerges 1987), the Biase response to these projects has been unenthusiastic. Since people rarely use the bus stops, some villagers have converted them to other uses. Many are used for storing building materials such as sand, bags of cement, and lumber; some serve as kiosks; a few more are used as lounging areas.

A glimmer of hope seemed to appear for rural Nigerians in 1987 when the wife of Nigeria's president introduced what was described as the Better Life Program for Rural Women (BLP), which aimed to improve the living conditions of rural women. Incidentally, there was already a Directorate of Foods, Roads, and Rural Infrastructure (DFRRI) set up by the government to improve rural conditions, but the president's wife accused the directorate of having no program that specifically addressed the problems of rural women. This prompted her to found the Better Life program. Although I arrived in Biase three years after the program started, no one in our villages was familiar with BLP. *The African Guardian*, a Nigerian journal, reported in 1990 that many rural communities have never heard about the program, and that some others are disenchanted with its operation. At Ilado-Ado in Lagos state, one of the villages where officials claim the program started, the leader of the Market Women Association said she had no idea what the program meant, nor has a BLP scheme ever been established in her village. Mary Kanu, an official of Women in Nigeria (WIN), believes "some groups of urban women elite decided this program for themselves for the purpose of shoring up their own image" (1990: 21). Rather than transport rural women from the villages all the way to Lagos to participate in Better Life exhibitions, Kanu says it is better to "divert all the state funds they are using for this vanity show to concrete programs . . . [because] it is merely a show of eminent urban women mimicking the rural poor" (p. 22). WIN's deputy coordinator, Susannah Prince, predicts that "no future government would continue with this waste of public funds" (p. 24).

Disrespect for Rural People by State Agents

Such negative experiences made the relationship with state agents sometimes confrontational. They also demonstrated how rural ways of life conflict with the expectations of the state. There were incessant conflicts between the Biase and state officials: policemen illegally raided Biase villages and confiscated personal belongings, government forestry officers prevented the Biase from using forest resources, and health inspectors prosecuted them for cultivating gardens too close to their homes.

Three months after I arrived in Biase, the health inspector assigned to the Biase sent notices to the villages informing them of a forthcoming inspection. When he arrived in Emomoro village, he went to Onun Emomoro's house, where other Inun were waiting to meet him. He gave the Inun a list of violations he had noticed as he walked through the village. There were fallen trees across pathways, ponds in which mosquitoes bred, garbage heaps collected in front of homes, overgrown brush which could harbor snakes, and some food crops planted too close to homes. The Inun explained that such circumstances were all part of the way we live: trees serve as temporary bridges when the village is flooded, the ponds result from digging up earth for house construction, garbage heaps are frequently burned so the smoke will help control mosquitoes, and plantains are always grown in home gardens (plate 6). The health officer insisted that the village was in violation of the law and imposed a fine of two thousand Naira. Sensing their relative powerlessness and judging that the official wanted money, the Inun began negotiating to pay him off. After a lengthy negotiation with one of the Inun, the officer settled for sixty Naira. The high point of this incident did not occur until two weeks later, when the same health inspector returned unannounced and reinstated charges for the violations for which we had previously been fined. This visit infuriated the villagers and provoked an angry philosophical rebuke from the oldest man in Agwagune, Onun Jonah Aquah: "I do not walk around with a cane out of pride, but because of the several trips I have made for government. I was customary court president before your people in Ukwa started going to school. My legs were stronger than yours when I was your age, they'll still be stronger than yours when you're my age. The fact that you see us living inside a hole, does not mean we do not feel the sunlight. We have tried to make you our friend. If you want to make us your enemy, you will find the Agwagune waiting for you." Onun Aquah saluted his fellow Inun, picked up his hat, and walked out, signaling the end of the meeting. The health officer quickly ended his visit and never made another visit to Agwagune while I was there. But the officer's visit was insignificant compared with what followed a month later.

Following a botched military coup in April 1990, policemen from a local station began searching homes in some Biase villages supposedly looking for escaped coup plotters. In Abini, Akpet, and Abanwan villages, they broke into homes, confiscated electronic equipment, beat people, and forced them to open their homes for inspection; many villagers who could not produce purchase receipts for household items lost such valuables as electric power generators, radio sets, and bicycles, because they could not pay to redeem them. When news of this event reached me in one of the villages, I was concerned about my research equipment and hurried back to my primary site in Emomoro village. The policemen had not yet arrived, but there was much excitement in the village. My field assistants were relieved to see me. They advised that we should carry all my equipment to the bush and hide it in a safe place until the excitement was over. Although I had receipts for every piece of equipment, I was uncertain about the prospect of confronting one of the officers; I did not want to learn by hindsight. I followed their advice and, with their help, carried my equipment about a hundred meters into the forest. Silently we cleared and dug up a spot, wrapped the items in cloth and paper coverings, and strewed leaves around the spot. For two days, my computer, cameras, and batteries remained in the bush under earth and dirt. By the beginning of the third day, news came that the policemen had been acting without authorization and that some of them had been arrested by officers from the state police headquarters.

Villagers were often harassed by state forestry officers enforcing state laws that govern the use of forest resources. In 1978 Nigeria's federal government passed the Land Use Decree, which vested all land and trees in state governments, shifting control away from traditional rulers (Swindell and Mamman 1990). This was clearly in conflict with customary patterns of resource use and allocation, in particular, the Agwagune system of tree ownership. A forestry agent told me he was once chased from the forest by a machete-wielding villager he had gone to arrest. To avoid such risks, some forestry agents, rather than go to the forest, simply walk around our villages inspecting dwellings and confiscating lumber they suspect may have come from forest trees. Villagers believe they are frequently bothered by agents of the state because they lack political power and are excluded from prominent city connections.

Conclusion

These various changes in the Biase internal political arrangements have weakened our traditional authority structures, making them less able to respond to pressures exerted by the new patterns of social interaction. This weakness has increased the vulnerability of the Biase in their relationships with neighbors, such as the Efik and the Igbo, and relegated them to a peripheral position in negotiations with the state. In Agwagune, we often measure our political marginality by comparing the growth of our villages with that of those of the Efik, with whom we historically shared political power. That power has shifted in favor of the Efik: the Cross River State capital is sited in the Efik city of Calabar; the political head of the Efik is a high-ranking member of the state council of chiefs, exerting political influence not only among the Efik but within the state legislature; and government-provided amenities in Calabar are more easily accessible to Efik villages than to other groups located further from the state capital. This has created political and economic advantages for the Efik over other regional groups in the state. The political and economic powerlessness of peripheral groups such as the Biase has increased.

EIGHT

Ideology

"I will make life difficult for the human race. A great famine is going to fall upon the land because you will search and not find me. The only one who gets your respect is yam although none of you lives on yams every day. If this slight is not reversed, the human race will be unable to feed itself much longer." After this warning, farmers from all over the world set aside a week in which they hold festivities to honor and appease the cassava crop.

Agwagune informants report that a woman who went to the farm one morning to harvest some cassava heard this speech by her cassava and was thus able to save the world.

I begin this chapter by relating the above incident to illustrate the coincidence of the real and mystical worlds of the Biase and the various strategies we use to maintain an equilibrium between them. We seek this balance because we must not only protect ourselves from the machinations of evil-intentioned neighbors but must also enhance survival opportunities in an uncertain environment. As traditional coping mechanisms continue to weaken, this need seems more urgent. The Biase explain the problems of culture on the basis of the tangible and mystical worlds. Often we explain the happenings around us by attributing them to our neglect of certain powers in the mystical world that reinforce the tangible world. Attributing causality to otherworldly powers is a dominant theme in the daily life of people in technologically less complex cultures such as the Biase, and people strongly respect the prime importance of the mystical world for their livelihood.

A month after I arrived in Biase, my Agwagune informants reported there were two men in Odumugom-Agwagune who employed magic as a means of increasing their success in fishing. They always caught more fish than anyone else fishing in the same area. In one season, the two men made so much money from their catch that they were able to change the roofing of their houses from thatch to corrugated iron sheets. One day I mentioned this to some men lounging in a lineage center. Many confirmed that the two men used magic and pointed to a sore on the foot of one of the men as evidence. They explained that to "buy" the magic, Ayamba had to accept the affliction of the sore in exchange for good fortune at fishing. The magic of Etan, the other man, was embodied in an old army coat he often wore while fishing. According to the story, Etan bought the magic from a medicine man in the Ibibio region of southeastern Nigeria. Since obtaining the magic he has had a series of successes at fishing. Another man, who insisted that the men had not used magic, got into an argument with the men at the center.

> Not a penny's worth of magic. You all are saying things you have no idea about. Iyam, listen to me before these people start damaging someone's reputation. I have been sitting here listening to all this foolish talk about Etan and Ayamba and magic. Lies, that's what you are hearing. After hearing all the rumors, I myself approached Etan for help in getting the magic. Listen to what Etan said to me. He asked me, do you want magic? I said yes. He said, very well, bring fifty Naira, two bottles of hot drinks, and a keg of palm wine. When I woke up the next morning, the first thing I did was to knock on Etan's door. We sat down and Etan poured some wine for me and poured some for himself. He poured again for me and poured for himself. Then he looked at me and asked me, "Odim, how many meri-meri lines do you use for fishing?" I replied that I used about ten lines like most of the other fishermen. He smiled and said, Odim if anybody tells you that I use magic for fishing, tell him that he is telling a lie. I am not saying that there is no magic for fishing, but this Etan sitting before you uses nothing like that. The only magic I use are my meri-meri lines. When people like you use ten lines, I use twenty or twenty-five; sometimes thirty.

The general opinion after Odim stormed angrily out of the center was that he was not telling the truth. I learned that in the same week Odim visited Etan he also started catching more fish than usual. Some said Odim's catch from a single fishing trip that week was enough to feed the whole of Egbisim for a couple of days. This was strong evidence that he had availed himself of the magic he sought.

When I visited Etan a few days later, I noticed that he not only used more meri-meri lines than others but also had a line of sixty nets stretched along strategic sections of the streams and the river. I accompanied Etan on two fishing trips to see what else he was doing besides investing more capital in his business. The more critical factor I observed was that Etan had about five thousand square feet of water to himself whenever he went to put his lines in the water. I found this curious, because in other fishing spots, there was an average of eleven men sharing an area of similar size. Etan explained that only a few people ever fished in the same area with him because they feared he would use his magic to direct fish away from them to himself.

In this chapter, I will examine the connection between the Biase belief system and their economic system and how conforming to the culture and challenges to it influence the contribution of individuals to provisioning the community. I will attempt to show how our symbols and ideologies influence the production decisions of Biase farmers.

The pervasive influence of ideology in cultures such as ours has prompted the important suggestion that the attitudes, beliefs, and worldview of farmers must be taken into account if others are to understand the choices they make (Nardi 1983: 698). This view has long been supported by case studies that report project failures resulting from inattentiveness to how traditional ideology influences the responses of the project beneficiaries. Many years ago, the Mossi of Burkina Faso failed to adopt a new farming technology introduced by French colonists who neglected the interaction between Mossi traditional ideology and their farming system (Hammond 1959). The eagerness of French officials to involve the Mossi in the Niger irrigation project isolated the Mossi from the supernatural aspect of their environment. Their economic life was directly linked to rituals performed in honor of the earth custodian, but their involvement in the project entailed being resettled in project villages away from the familiar environmental forces they perceived as relevant to their economic survival. Consequently, the settlers were dissatisfied with their new environment and adapted poorly to the attempt by the French colonial government to involve them in new agricultural techniques. These internal cultural factors (see also Grayzel 1986)[1] are important for understanding the Biase situation because isolated communities such as ours tend to explain and attempt to resolve current problems, including environmental crisis, on the basis of our traditional ideologies and indigenous coping mechanisms.

Subsistence and Ideology in Biase

The Biase believe in the efficacy of magic as a means of influencing environmental conditions affecting subsistence; reliance on magic is not

1. A similar case is reported about the FulBe of Mali, whose traditional culture is tied to their subsistence strategies (Grayzel 1986).

exclusive of the application of indigenous technology but is complementary to Biase ethnoecological knowledge. Magic helps the Biase interpret the variable capabilities of community members who perform too well or are unlucky on the basis of how their magic or technology responds to their desired goals. To understand this process, I use the concepts of "nature dependency," or an exclusive dependence on nature for farming success, and "magic dependency," in which success is attributed to magic. With nature dependency the Biase rely on the visible and invisible forces of nature, which are beyond their control, for the outcome of their investment in subsistence. The tangible elements, such as capital investment, amount of arable land, labor availability, crops, and environmental hazards, are all directed by the deity Erot Edok, the invisible force of nature associated with farming. Magic dependency is the attempt by the producer to control the visible and invisible forces of nature.

These conceptual terms do not classify separate domains of production, since nature and magic are inseparable in Biase production technology. I specify them only to clarify their complementarity and in the process explain the pivotal role of ideology in the subsistence strategies of the Biase. Their effect on economic growth depends on how much the Biase believe in their efficacy and on how they act or react in response to the observance and violations of rules by fellow villagers.

Nature dependency relies primarily on the nutritional content of the soil and whatever else crops get from the environment, some additional inputs such as fertilizers, and the mercy of our fertility deities. These natural production aids do not perform on their own but are coaxed into favorable disposition by the observance of rules, *idom,* that prescribe and regulate appropriate behavior toward community deities, *erot,* or other community persons. Erot rules are associated with occupational matters, particularly farming and fishing; they are seasonal in their demand on adherents. Occurring mostly at the beginning or the end of the farming season, they specify what we must do or avoid in order to maintain the goodwill of the fertility deity. Some deity-centered proscriptions are: (1) Communities must perform fertility rituals before land distribution begins. (2) People must wait for everyone else to complete bush clearing before setting fires. (3) Harvesting may not start until the yams at the fertility deity's shrine have been harvested. (4) Men are prohibited from fishing in ponds that are designated for women. (5) Women are prohibited from wearing clothes when fishing some of the sacred ponds. (6) There is a prohibition against processing certain kinds of food inside the village. The processing must be done on the farm or in the bush. (7) Some villages forbid walking around the village at night with a lighted cigarette. (8) We are prohibited from engaging in sexual intercourse in the bush. (9) In Agwagune, fighting or quarreling at night is prohibited. (10) We are prohibited from relieving ourselves in front of the erot shrine.

Violating these proscriptions affects the quality of the next harvest and the capability of the entire community to provide sustenance for itself. It is easy for us to honor most of these proscriptions because their seasonality (following the farming cycle) minimizes the conditions for violations. Individuals do not constantly face the need to conform to them; besides, abiding by them is part of growing up in Biase. When there is a violation, however, the offenders are informed of the offense by attendants at the village shrines and told what they must do to be cleansed of the offense. In many cases, compliance is prompt. Anyone who fails to comply is barred from associating with people outside his household or from entering the bush to do any manner of work. Since such restrictions threaten the survival of any rural person, there are few cases of noncompliance.

Violating erot beliefs influences farming activities if the consequences of violation are perceived to directly affect the entire community or sections of it—that is, the village or lineage, men or women, both men and women, the individual, or the community. If the community is affected, farming activities are delayed until the offender performs the necessary cleansing ritual. During bush clearing in the year of my research, bush fire crossed from the plot of a woman of a neighboring village into Agwagune farm territory, partly burning an already-slashed bush area before the Agwagune were ready to set the fire. Farmwork stopped temporarily as both villages attempted to sort out the matter. Although a delegation arrived from the woman's village three weeks later to cleanse the offense, some farmers gave up their plots in the violated area, believing the plots would no longer do well. A few others took farm plots in other fields less suited to the crop they had intended to plant. Sometimes the offender does not have the means to perform the ritual, in which case the offender's lineage or family has to raise the money.

Violating farming proscriptions does not often result in work stoppage. Some offenders just do not have sufficient resources to perform the ritual cleansing. In these cases, priests perform rituals to forestall an evil outcome. During preparations for the beginning of farming in the year of my research, names of persons from various lineages were mentioned at the village meeting who had committed various violations in the past year. Priests warned of serious consequences if the rites were not performed within seven days. Most of the violations had not been ritually cleansed when I left the field many months later.

Unlike with deity-related proscriptions, the regular processes of daily interaction provide numerous circumstances for violating human-related proscriptions. Some such prohibitions include: (1) stepping over one's occupational tools; (2) passing behind a person while handling fire or hot coal; (3) passing or receiving any item with the left hand; (4) passing by an individual on his right-hand side; (5) greeting or speaking with someone when he or she is performing a ritual; (6) sitting at ritual spots reserved for specific individuals; (7) spitting on someone; (8) exchanging money at night or going to a neighbor's house at night to demand the payment of a debt.

Violating any of these proscriptions or refusing to perform restitution rituals may seriously affect the fishing or farming success of either the violator or the victim of the violation. Although all offenses are perceived as having negative consequences for the entire village, we view violations against humans more seriously than those against the deities. People who feel violated take immediate steps to exact retribution. Offenders are compelled to comply in order to avoid blame for whatever misfortune the offended party may later suffer. Although these ritual violations may have minimal effect on a community's agriculture, the offended individual may perpetually evoke them to explain subsequent misfortunes in other aspects of her life.

Nevertheless, people tend to show greater caution in observing proscriptions related to deities than for those affecting fellow individuals. Of the 60 percent of respondents who admitted ever violating any proscriptions, over 80 percent had violated more individual-centered proscriptions than deity-centered ones. Some say it is safer to offend other humans than to offend spirits because the consequences of violating rules related to deities are unpredictable. Various forms of unanticipated illness and misfortunes may be visited upon the community by an offended deity. Besides, collective pressure on offenders to prevent communal suffering is always enough to discourage acts against deities. If individuals are offended, people can easily negotiate the mildness or severity of their offenses and the complexity or simplicity of the cleansing rituals involved. Violations against deities carry standard ritual specifications and follow known procedures that are inflexible and rarely circumvented. This inflexibility explains why many violations have not been cleansed, since most offenders are those least able to meet purification obligations.

Erot Edok—Festival of the New Yam

Good-quality harvests and abundant food result from honoring the various proscriptive rules. The successful provision of sustenance, which we attribute to the erot deity, is annually celebrated by the entire community in a day of intense festive activities featuring visits, dances, and sharing among community members and visitors.

Our major harvest ritual, Erot Edok, is exclusively for honoring the yam deity and is performed before we start harvesting our yams. Early in the farming season, priests and assistants of the shrine clear a piece of land in front of the Erot Edok shrine, make seven earth mounds, and put yams in them. Only after this ritual do lineages divide up farmlands for their members. The yam deity is honored again in the month of August or September shortly before harvest begins. Late in the evening of a designated day, elderly women capture an unmarried maiden (who does not have to be a virgin) suggested by the deity and take her to the erot compound. They keep her there for the night and prepare her for the following day's ceremony. The edok maiden comes from the kingship lineages, Emo Imo or Emo Edodi. As the time appraches for capturing the edok maiden, teenaged girls from those lineages sometimes disappear for days in order to avoid capture. It is said that Erot Edok always finds an appropriate maiden. One year the desired maiden was away in the secondary school far from Agwagune. Erot Edok priests were in the process of convincing Erot Edok to chose another maiden when the chosen maiden unexpectedly arrived in the village from her school and was immediately captured by the Erot Edok priests. No bad consequences are known to result from serving as the Erot Edok maiden; on the contrary, no woman who was an Edok maiden has ever gone without giving birth to many healthy children. On the morning following her capture, the maiden appears all painted in white chalk (plate 28), carrying the edok basket delicately balanced on her head, and is led around the villages by Erot Edok priests. The village people run behind her and the priests sing Erot Edok praise songs that signify the abundance of food provided by the yam deity. To mark this occasion, other festival plays are performed by groups of friends, age sets, or other men's and women's secular associations.

> *Nam mi nji efa, No are-e*
> (I wish to eat cassava, but you say)
> *Nam mi ndup odot, No are-e*
> (I wish to lick vegetable sauce, but you say)

We start harvesting yams about a month after the celebration of the new yam festival and continue through the month of December, when men construct their yam barns. All through the season, the harvesting of other crops such as plantains, cassava, cocoyams, vegetables, and other cultigens continues without any ceremonies. We believe that anyone who plants or harvests yams before the rituals are performed will cause the harvest for that year to be poor, resulting in possible starvation for the community. But as with other Biase proscriptive rules, people consistently violate this custom. Informants reported that some people have harvested yams and taken them to their homes surreptitiously before the Erot Edok ceremony. They are usually not called to account if no one sees them enter the village with the yams.

This was an apparent contradiction between preharvest rules and the community's willingness to overlook infractions. If harvesting yams before the festival portended bad luck for the community, why were people risking the community's well-being? And why was the community aware of and tolerant of this violation? Many of our people did not share this concern. Some told me that bringing new yams into the village before the festival does not constitute a violation unless the yams are exposed to public view. Such exposure of the new yams signifies boastfulness by the individual who is symbolically announcing her ability to feed herself and the community without the deity's help. As the Erot Edok priest of Abini explained:

> All that spirits want us to do is be respectful, not perfect. Do you think I'm an Erot Edok priest because I'm perfect? Erot Edok does not watch your back to know if you are doing wrong or doing good. That job is my job. If you do something, I speak with you and say, Iyam, you have fallen short of this or that. This is what you should give to Erot. If you say you are too big to do what I tell you, then Erot will take over your case, and that is not small money. Some people come back begging, but I say, my hands are no longer there.

Violators rarely let infractions go beyond the willingness of the priest to intercede because of the maze of sanctions and sacrifices required to appease local deities. But the prohibition against exposing new yams is variously interpreted, depending on whether the whole yam or only part of it is exposed. I occasionally saw the crop partly exposed in people's carrying baskets. Some farmers simply carry new yams on their heads and drape a small leaf over them to symbolize concealment without incurring sanctions. One day as I lounged with some men at a lineage center, a farmer carrying a barely covered yam passed by and exchanged greetings with us. When I interrupted the conversation to ask those at the square why the yam was not properly covered, someone responded: "What yam? Did anyone here see a new yam that was not covered? If you saw one then you must be a spirit because all the new yams are still in the ground. If you are hungry for a new yam you can follow me to the farm tomorrow where we can safely eat it in my farm shed without announcing it. Otherwise wait until next month when Erot Edok shows the new yams to the village." I seemed to be the only one who perceived a contradiction in this practice, because throughout the August of my research year, people occasionally consumed new yams in their homes before the performance of the festival late in September. It was improper, however, to announce that you were doing so. When I visited farms, some farmers had already harvested yams which they kept in farm sheds. After Erot Edok, the yams would go into barns to be stored until they were sold or consumed or to be kept until the next cropping.

On the festival day it became evident that people had been harvesting yams since long before the festival. Even before festivities began, people in many homes spent the morning preparing large amounts of pounded yam meal for friends and guests visiting for the occasion. Although people often said they harvested their yams on the festival morning, this was rarely the case; most people stayed home preparing for the celebration.

Our traditional beliefs tend to be directives on how to respond to natural forces that influence our means of making a living rather than sets of restrictive rules. Consequently, the observable contradictions of accepted beliefs do not violate those directives but function as strategies for surviving within them.

Magic dependency is the farmer's belief that a good harvest depends not only on the gifts of nature and the goodwill of the deities but also on controlling the evil designs of fellow farmers and influencing and forcing the goodwill of other supernatural forces. I was told that one could lose those gifts through the cumulative effects of violations or by the magical means used by some farmers to attract the fortune of others to themselves. After experiencing a bad season, a farmer may attribute his misfortune to the intentions of another person and acquire magic as a means of offense or defense in the future. Magic is also used to induce a good harvest. This is better understood in terms of the more modern farmer's use of light technology such as fertilizer, new plant varieties, or some modern form of inducing growth. For the Biase farmer who uses magic, the input assumes a less concrete form and becomes merely symbolic. Its efficacy, however, is no less potent.

Magic for inducing increased food production may be either displayed publicly in the farm plots of the magic's owners or used privately without the knowledge of others. Private magic is said to be employed selfishly by individuals who interpret success in terms of the misfortune of their neighbors. Usually there is no material indication that such magic exists, but a rumor is easily believed when the object of the rumor experiences greater success than others. My Abini informants told me the story of Ebi, who often became ill at the beginning of the farming season. Because of his ill health, people would go and help clear and till his farm and also plant his crops. As soon as his yams were safely in the heaps, he would suddenly become well again. At harvest time, Ebi's yams did better than the yams of all the other farmers. After two farming seasons, people suspected that Ebi was using magic to divert the good luck of all those helping him to himself. In subsequent seasons, when people stopped helping him, his yams no longer did better than those of other farmers, and he was no longer ill at the beginning of the farming season. There were numerous such cases of individuals seeking to maximize personal goals rather than the community's. This is often a source of discord between owners of magic and community people who feel they were unwittingly contributing to someone else's success. Although no sanctions exist against the use of magic, villagers suddenly get distrustful of anyone suspected to be employing private magic, because whatever success the individual derives is believed to result from the misfortune of a neighbor.

As we sat in the village square one morning, a man told me the story of a man who traded his brother's life for greater success at fishing. Ukam took a photograph of his brother to a distant village known for its medicine men. When he brought out the photograph and informed the medicine man that he wished to trade his brother's life for economic success, the medicine man told him that his brother's spirit was too strong for him to trade. He accepted the medicine man's option that he trade both his brother's life and his own for success and wealth that would last seven years. The brother's death would come in a few months, while he would die at the end of the seventh year of accumulating wealth. Shortly after the man returned from that trip, his brother drowned in the river.

Ekpezu went fishing one evening with other fishermen.

> Suddenly, Ekpezu's canoe started pulling away as if it was being driven by a motor. All of us were still putting our fishing lines in the water that evening. My friend Eko and my brother Uso were closer to him. I did not know what was happening until I heard Eko scream. When I looked up it was as if I were dreaming. Ekpezu was slumped down in his canoe and his paddle was inside the canoe. This thing kept pulling his canoe away. Then suddenly the front of his canoe dipped in the water and we didn't see anything after that. The three of us then turned our canoes and started going in that direction. My brother and I went in and searched for a long time but we did not see anything. We did not even see the canoe. That was when I knew that this was not an ordinary event. Other men returning from fishing joined us and we searched the whole area; there was no sign of Ekpezu or his canoe. The next day, we saw his canoe tied to the bank of the river; his fishing clothes were neatly folded and placed on one of the rungs. Three days later, we discovered his body right where his canoe first disappeared.

People associated this tragedy with the success Ukam was having at fishing, particularly after he made some money and left the village to start trading in clothes. Twelve years after the incident, some report he is alive and doing well at his new occupation. I was told that he is staying away from Biase to thwart the effect of his agreement with the medicine man.

Another form of magic used to enhance survival chances is more public and more generally accepted. My Agwagune informants said that yam farmers in the neighboring town of Abanwan always have a good harvest and maintain large yam barns holding between five hundred and one thousand yams at a time. The reason given was that every Abanwan farmer has yam magic that helps bring a good harvest.

My follow-up investigation confirmed the large yam barns some Abanwan farmers owned. Some barns were so large you could easily lose your way in them. Inside some of the barns were small shrines that complemented the efforts of the farmer to attain a good harvest. Abanwan men also own and worship a yam deity and believe, as the Agwagune do, that their harvest depends on the goodwill of the deity. But to help the deity in his work, the farmer must seek the cooperation of lesser spirits who help the fertility deity convey all the fortune provided for every farmer. Failure to maintain individual shrines complementary to the communal shrine usually makes the difference between one farmer's abundant harvest and the paltry harvest of another. As a testimony to the truth of this belief, such individual shrines are usually found in the larger yam barns. But individual shrines are also present on smaller farms. I was told that some of the shrines were meant merely to scare off potential pilferers rather than to increase production.

The Abanwan yam farmer takes his yam cultivation seriously and invests capital not only in acquiring magic but also in appealing to the logic of scientifically increasing his production. While Agwagune farmers invest an average of two thousand Naira in yams, Abanwan yam farmers invest an average of four thousand Naira. The major incentive for the Abanwan farmer is the community's location close to such commercial Igbo towns as Afikpo, from which traders come in large numbers to buy yams for the large Igbo population. Although Abanwan also has the problem of poor roads, it has a fairly usable dry season road which helps in the transportation of food crops usually harvested at that time of the year. Additionally, the town's yam barns are strategically located close to a good vehicle access way. Vehicles drive right up to some barns where farmers wait to sell and load their crop for transportation to other Nigerian towns. Generally farmers tend to believe that one's prospect for a good yam harvest is as dependent on how much capital is available for labor and yam seedlings as on magic.

Even farmers who did not own individual shrines believed that such shrines improved the opportunities of other farmers. I visited the farm of a man who kept a magnificent yam barn but did not appear to have a personal shrine anywhere around. He explained: "Only lazy people need magic to help them do their work. Some years I have good harvest, some years it is bad. But I don't go to those people and say, please give me food. When we have a flood, everybody's harvest is bad even if their magic is as tall as this tree.

He then pulled up his shirtsleeve to reveal a copper armband and explained that it was to protect himself and his crops from jealous neighbors. So in addition to using magic for making crops grow, some farmers also kept offensive or defensive magic to cope with the uncertainty surrounding the source of abundance and scarcity. Although a yam farmer may not be certain which investment results in increased production, he has no doubt that one must complement the other.

Finance is the limiting factor to increasing production by either magical or scientific means. So as in every business situation, the variable financial resources of producers affect the final result. Only a few can afford to buy adequate fishing tools or enough yam seedlings, as well as to hire sufficient labor. Some farmers spend as much for personal magic as for the total cost of yam seedlings, labor, and transport. However, magic is a onetime investment, only needing to be maintained through inexpensive seasonal rituals. But purchasing yam seedlings, new fishing lines and nets, and hiring labor require a significant yearly capital investment.

The expectation individuals have of the consequences of violating communal proscriptions is sometimes pivotal in their decisions to withhold farm inputs that would improve production. Their perception is that violations invariably lead to unavoidable consequences whose effects may be mediated but not averted. Individuals expecting to suffer repercussions for violations of proscriptive rules become despondent and unwilling to risk much capital and effort to reverse an inevitable outcome. In Agwagune, a woman told me about her attempt to use manufactured fertilizer on her vegetable garden. She followed the instructions of the product's seller and carefully applied the fertilizer to her plot. A few days later, she noticed that all her vegetables had turned brown:

> At first I did not know what happened. I kept saying to myself that someone has put something in my farm plot so that I will not sell vegetables this year as other women will be doing. If you saw my vegetables, you would think that someone poured hot water on all of them. When I went home and told my husband, he screamed and said it was his *mfam* (magic) that had caused it. There are things you cannot take onto the farm if you are using mfam because you will spoil your crops. So his mfam punished me for my action. Since then, I have fertilized my crops only with ashes from my kitchen and my vegetables look just like the bottom of a baby born today.

Sometimes even when farmers are aware of other scientific methods to increase production, they may avoid using them because of proscriptions related to magic. For example, the Agwagune prohibit people returning from fishing from walking through farm plots unless they have first washed off the mud from the pond. Since the route to some of the fishponds passes by people's farms, people stop by ponds on the way home to bathe before passing through farms on the return route. There was a case in which a woman complained that a certain boy walked through her farm without first bathing. Because of this incident she stopped going to work on her farm since she was certain her crops would die no matter how much work she put into saving them. Subsequently, she stopped watering her crops, her farm overgrew with weeds, and the crops died. The woman took her case to the Abu, which ruled in her favor and made the boy's mother pay twenty Naira as the cost of the damaged crops. The boy's family was also required to perform purification rituals to appease Erot Edok.

The cases reported here illustrate how, among the Biase, nature and magic contribute in various degrees to the total amount of food available for consumption. Fishermen who believed their colleagues were using magic to increase their catch attempted to find ways to replicate their success. In so doing, some discovered that the way to increase production was to manage their businesses better; the result was increased production. This is also apparent in the cases of farmers whose dependence on public magic was complementary to scientifically based input (labor, time, quantity of seedlings, amount of land used). Success depended on using both factors.

Where magic is not used, the farmer depends on the natural environment and people's conformity to proscriptive rules to get a good harvest. If farms are not damaged by excessive flooding, Biase farmers often have relatively good harvests. For some farmers, however, production may be affected by their responses to violations of behavioral rules. Production may decrease because a farmer stops working because he expects disaster; farmers are likely to produce less if they believe it is futile to try to increase production. Farmwork stops as farmers attempt to sort out the rules and food set aside for domestic consumption is used to fulfill purification rituals.

Over the past two decades, many of the rules governing the treatment of violators have been relaxed. An elderly informant attributed this change to returning Biase migrants. She traced the situation to some Biase men who served in the army during the Nigerian civil war. Many who returned were indifferent to traditional beliefs after confronting the ordeal of war. "Today you may see a child calling his grandfather a witch right in the marketplace. People keep tempting and insulting you so that if they have a headache they can say, it is Abei who is trying to kill me with her Inyono. Next thing you know, someone shoots and kills you or the police take you away to Akamkpa or Calabar." In 1970 a village Onun was beaten by a returned emigrant who accused the elder of using magic to cause him to be discharged from the army at the end of the civil war. Biase villagers suddenly faced a new threat from young men out to rid the villages of "old witches." There were rarely communal sanctions against such actions because people felt helpless against former soldiers suspected to be in possession of dangerous arms. This passivity helped accelerate the declining status of community elders. Additionally, returning emigrants from Biase were often ignorant of local proscriptions, and violations by non local people are generally treated less seriously or even ignored. The cumulative effect of these factors, coupled with state laws, led to relaxing sanctions against actions of people that violated the worldview of the Biase.

The Declining Role of Deities in Biase

Accompanying the loss of reverence for institutional proscriptions was a new conception of the relationship between the Biase and their deities. My Agwagune and Abini informants told me that some important deities have left their communities for more friendly locales. Their absence is frequently invoked to explain the seriousness of some of the environmental disasters the Biase have been experiencing, as well as the inability of the Biase to improve their condition.

In Emomoro, I accompanied members of Ase Egwa lineage to Ibe-e-man, a sacred pond that once served as the home of a fertility deity. I was told that this deity occasionally visited her old habitat and that a few people had sighted her on relatively quiet days. We had to exercise caution, however, because some of the pythons who often served the deity were still there. Ibe-e-man is located equidistant from all the Emomoro farming areas in this zone and provides drinking water for farmers, who either collect water from it on the way to their farms or make a trip back from their farms to get drinking water during hot afternoons. I was told that Ibe-e-man is never without water even when the whole earth is dry.

Traveling along a farm trail, we made a detour down a narrow trail that dropped sharply to about twenty-five feet below the farm trail. Around a bend, we came to a dark grove under tall oak trees. As I pushed forward, one of the men quickly pulled me back, and cautioned that we were at Ibe-e-man. Quietly, another man crept forward on hands and knees, stopped and peered into the darkness, and motioned for us to follow. We walked a few feet before stopping at the edge of a small pond covered with dried leaves, broken twigs, and a variety of forest debris. Some people have reported sighting the Ibe-e-man goddess. One report said a woman went there to collect water on a hot afternoon and briefly saw a woman sitting by the pond. In her fright, she was unable to look at the sight and she quickly turned and ran. When her fellow farmers accompanied her back to the place, the goddess had disappeared. According to popular accounts, the goddess's body is painted all over with what looks like white chalk. In the hope of seeing the goddess or one of her python servants, we approached Ibe-e-man cautiously and respectfully. On this day she was not home.

Some informants believe the goddess fled to another community because the Agwagune neglected to perform cleansing rituals, and stopped caring for her dwelling. Previously, before the farming season began, some age sets would be sent to clean away dirt from the pond and clear the brush around the area, making it more habitable for the goddess and easier for farmers to get water from. No such activity has been organized in recent years. A few people attribute the increased incidence of environmental problems in Biase to the flight of other important deities who feel neglected by the villages they once supported.

Biase proscriptions have important relations with other aspects of the society. By standardizing the times people start specific farm tasks, we ensure that the labor supply is equally distributed among the farmers. There may still be differences. Those able to afford greater capital investment or to hire extra labor are able to farm more land and work it faster than others. But the custom provides the feeling that no one has an unequal advantage. For this reason we tend to be suspicious of people who suddenly excel in occupations in which all are perceived to be equally competent.

In the past thirty years, many ritual observances have been ignored, discontinued, or even challenged in the state court. Traditional beliefs and practices which earlier served to sustain our economies have died out. People differ in their perceptions of the efficacy of traditional beliefs. Although we still honor communal deities in annual and seasonal festivals, some younger community members seem no longer convinced that ignoring those festivals and the rituals associated with them would result in a communal disaster. Young community members no longer see themselves as defenseless against supernatural forces or powerless against the mystical powers of our elders. But many community members still rely to a great extent on supernatural forces in farming, fishing, and expectations for the future. This paradox affects the responses of community members to the challenges of everyday life and the demands of our traditional beliefs and practices.

Although people have violated rules without suffering any immediately perceptible consequences, we believe that retribution is imminent unless the offenders appease those forces that control the ethereal world of beliefs and magic. We believe that to enhance the opportunities to produce adequate food, we must strengthen indigenous technology with forces and resources that appeal to the objects of our senses, such as magic and ritual practices, as well as to the objects of science, such as adequate labor and finance. Consequently, the symbolic elements of magic rituals, which are too often perceived as irrational, are as relevant to economic success as the concrete elements of scientific input or "rational" production decisions. We misinterpret the Biase farmers' perspectives when we separate these two forces by de-emphasizing the nontechnological aspects of production.

Conclusion

The weakness in our cultural beliefs and the loss of our fear of retribution do not interfere with our cooperative ethic, which is manifest more at the level of nondomestic than of domestic relations. Our religious beliefs still constitute an important basis for cooperation at a time when other components of our culture have been severely weakened. Indeed, our ritual practices extend the cooperative ethic by intensifying the level of interaction between community members, fostering social relations with spectators from neighboring groups, and maintaining our link with the mystical world. As I explained in the case of the Abu society, the performance of religious rituals is closely linked with the age set system. Members of a set are assigned the responsibility of supplying a particular item or performing a specific part of a ritual. One man may have to provide the goat, a second the palm wine, a third will go to the forest to get herbs, the women of the same set may supply the food, and another set may do the cooking. The Onun of each set coordinates the activities of that set. Communal participation has always been a part of our ritual performances, to the extent that the community does not encroach upon or probe the mystery that resides in the realm of ritual priests. Through ritual performances we regenerate our weakened bonds with our spirits and ancestors, although without the fear that formerly surrounded those performances. The various assignments bond participants within and across age sets, our completion of the ritual provides the faith we need to understand and interpret the world, and the result is the assurance that our wishes will be fulfilled.

Even in their relative powerlessness, our secular and mystical associations still perform important social functions such as mediating conflicts and regulating aspects of social behavior. Thus the associations foster social harmony and provide the medium for cooperation. Abu initiation rituals have lost their dangerousness but not their mystery. This development is positive because it broadens the level of participation by community members without impairing our admiration for Abu. On festive occasions, some of our women living in the cities sometimes perform Abu dances and sing Abu songs in violation of the Abu code but far away from the sanctions and reprimands that would be imposed back in the village. This seems to underline the weakness of the Abu society since formerly these women would have been killed. Under our present circumstances, women's discovery that they can steal our dance steps and sing our songs without losing their fertility or being punished by mystical powers serves only to humanize the society. The mystery in the ritual performances still authenticates the sense of male pride that Abu evokes. This mystery is shared by only a few members of Abu but is revered by all.

The prescriptions of the Christian church have also had some influence on our belief systems because many of us are now nominally Christian. Although the Christian ethical code is not different from the rules of conduct that have always been followed by the Biase, we have modified some aspects of our ritual practices to accommodate converts to Christianity.

One of the lasting effects of Christianity is that the Biase try not to do any manner of work on Sundays. When the market day falls on a Sunday the market is convened the next day to allow the Biase of various Christian faiths to practice their beliefs. Christmas is a festive occasion that is much anticipated by the Biase. Husbands buy their wives new clothes, shoes, and head ties; all Biase outside the community come home to visit friends and family and renew old acquaintances; men, women, and children dress in festive clothes and sing and dance around the communities; young men don their masquerade costumes and harass women and uninitiated men; the fragrance of fresh fish stew fills the air around every home, and food and drinks are exchanged in an atmosphere of continuous feasting that is sustained well into the new year.

Soon after the Church of Scotland Mission arrived in Agwagune in

1876, the Biase adopted biblical names and gave them to their children as a way of affirming their conversion and convincing the missionary fathers of their Christianity. A few years ago, the Agwagune caucus in Calabar addressed the problem that we were losing our local names to foreign names. In a series of meetings with Agwagune Inun, the caucus advised them to reverse this trend by giving their children only Biase names. Only a few parents today still give their children non-Biase names.

More than in any other social group, it is in our women's association that the cooperative ethic is most clearly marked. Besides serving as the primary social and economic support of our communities, our women organize locally and extend their cooperation to women outside their communities. In the process, they pick up the pieces of our community's broken hoe, reconstruct it within our reconfigured cultural system, and encourage our men to reassert their enterpreneurial spirit, if not their glory, in the hope that we will someday improve our situation.

NINE

Implications for Anthropology

Your Excellencies, the Military Governor of the former South Eastern State came here in 1970 when Agwagune suffered a major flood disaster. When he saw the problems, he promised to set up a model village in Agwagune. The project never saw the light of day. Since then, our farms and homes have been damaged by yearly flooding. One of the most serious floodings happened just last year. Government officials visited our community to assess the extent of the disaster. We were made to contribute money to the government to help the effort to rehabilitate us. But until now nothing has happened, and the affected persons have still not been rehabilitated. Agwagune is suffering from erosion, and many of our houses have fallen into the river.

We lack an all-season road to Agwagune because the only Trunk B road, the Adim-Agwagune road constructed by the colonial government in 1946, has suffered serious neglect since its construction. There has been no improvement in the condition of the road, nor has any effort been made to make it an all-season road or usable in the dry season. You may have seen that this road is impossible to use in the rainy season and difficult to use in the dry season. The importance of this road lies in the fact that it serves as the major link to the rest of the state for communities across the river and communities surrounding Agwagune.

We have neither a good water supply nor a borehole. We now drink water from the river and dirty ponds. We learned that the state government had long ago approved the construction of boreholes for Agwagune. Government officials came to survey the area and actually started construction two years ago. Surprisingly, the job was abandoned midway and the officials have not returned.

(Excerpt from an address of welcome presented by the Agwagune community in Biase Local Government Area to the military governor of Cross River State, Lt. Colonel Ernest Kizito Attah, and the deputy governor of Cross River State, Professor Solomon O. Unoh, on December 29, 1990).

Toward the end of my research, a landmark event occurred in Biase. Against all odds, the military government announced on August 27 that Solomon Ogbodim Unoh, the university professor who headed the Agwagune caucus in Calabar, was the new deputy governor of our state, the Cross River. Although Unoh was from Agwagune, the entire Biase group was elated by the news. It was an unimaginable event in the history of a Nigerian minority group. For the Agwagune specifically, it was an event that suddenly unfolded miles of dreams before us, seemed to provide the capability for fulfilling those dreams, and nourished in us the hope that our unexpected arrival at the corridors of power would unlock all the doors that had remained closed to us. Finally we were going to be ushered into government's inner sanctuaries where other Nigerian groups have for several years exclusively shared what is fondly referred to as the "national cake." Four months after his appointment, Unoh accompanied Ernest Kizito Attah, the Cross River State military governor, on a visit to Agwagune. After maneuvering the treacherous 9 ¼-kilometer route, what was left of the governor's entourage arrived in Odumugom. The trip gave Attah a firsthand experience of the dismal conditions in which most rural Nigerians live. The address above is part of the shopping list we presented to the governor.

Rural communities persistently ask their governments for provision of those amenities that would enable them to replicate the attractive ways of life of more advantaged groups. While the social relevance of such requests and even their potential for cultural destruction are still a central concern among scholars interested in the preservation of traditional cultures, these concerns figure very little in the demands of rural people such as the Biase for change. The question at this point is not whether fulfilling these wants would affect the ways of life of rural societies adversely but how to meet their demands so these communities can best control their own development agendas. Success is always more likely if development agencies begin by paying attention to local preferences rather than insisting on enforcing their own "good sense."

To Be or Not to Be Involved

Since the 1970s, anthropologists have continued to advocate the involvement of target communities in structuring change by being more sensitive to local conditions: "Projects had to be socially relevant, to be culturally appropriate, and to involve their direct beneficiaries in a significant fashion" (Escobar 1991: 663). In consonance with this new focus, anthropologists in development have since reversed the discipline's earlier nonadvocacy stance popular with British social anthropologists. Lucy Mair and Raymond Firth, for instance, "insisted on neutrality and technical rigor in conducting applied policy-related studies. It was considered 'unscientific' to pass judgment on official policies towards native peoples" (Wright 1988: 370). Almost thirty years ago, the Mexican anthropologist Guillermo Batalla (1966) faulted the dogma of cultural relativism, which he perceived in the works of anthropologists such as Edwin Smith, who felt that "if anthropology is to judge and guide it must have a conception of what constitutes the perfect society; and since it is debarred from having ideals it cannot judge, cannot guide, and cannot talk about progress" (Smith, cited in Batalla 1966: 90). The prevailing belief in anthropology that cultural practices are relative and adaptive does little to help rural inhabitants struggling to taste a bit of city life, particularly when the economic behavior of rulers and urban dwellers is made to seem so attractive (Hart 1979). Wright (1988) suggests that the underlying reason for this posture was the power asymmetry between anthropologists and colonial bureaucracies which discouraged aggressively interventionist viewpoints by scholars. This dilemma—"to be or not to be involved"—according to Escobar (1991: 668) has since been resolved in favor of intervention.

Unsuccessful development work is often blamed on the unwillingness of traditional communities to accept change. This often implies that rural people act with unanimity in matters affecting group survival. This opinion has been used to support a hypothesis of collective inertia among people in developing economies. (For example, see A. B. Higgins, cited in Ramirez-Faria 1991.) It has also in some cases become the basis for predicting how traditional communities would respond to various types of intervention. Less cautiously, this perspective has been conceptualized as evidence of the disinclination of people in less-complex cultures to be achievement-oriented (in contrast, of course, to the purported achievement orientation of those in the West). Ramirez-Faria (1991: 116) criticizes the ethnocentric inference that non-Western peoples will "develop" only when the Western achievement mystique is instilled in them. He recalls Higgins's opinion that, unlike technologically more advanced societies, peasant societies are characterized by the need of people to please their fellow members, to feel inferior to someone, and to have ideas and attitudes approved by persons they regard as superior, rather than being achievement oriented. These are indeed strange traits. In the African rural communities that I am familiar with, we are not inclined to please fellow members as a way of gaining approval or sanctioning their superiority. The cooperative ethic is present, but not any more so than in Western cultures. I reported in an earlier chapter that Ukpabi would not go to Calabar to pursue his business interests without the commitment of fellow community members. Ukpabi's decision was clearly a self-interested judgment rather than an attempt to please his fellow Abini.

Applied to understanding the complexity of the farmer's decisions, Higgins's grand misconception clustered the behavioral variations among rural communities into "types" (cf. T. Parsons), paving the way for the formulation of overarching ideas explaining the behavior of non-Western peoples. This formulation ignores individuals who pattern their activities differently and pass on those behavioral differences to others. In anthropology, it encouraged a theoretical orientation which as Goldschmidt (1990) suggests, was inattentive to cultural variations. While it explained the manifest arrangement of societies, it buried the essential role of individual actors amidst a series of metaphors. Thus, George Foster (1965) would assert in his "idea of limited good" that the traditional farmer perceives himself as powerless to increase resources through his own effort because doing so would jeopardize the well-being of his fellows. As a subsistence member of a moral community, she or he is said to shy away from adopting unfamiliar or potentially risky farming practices. Goran Hyden (1980) talks of an "economy of affection" rooted in the values of reciprocity and emphasizes how strongly rural economies resist capitalist values and incentives; peasants are strong with a negative power of resistance but helpless to expand production (Barker 1982). Goldschmidt (1990) blames this misplaced emphasis on culture theory, a theory of culture that neglects the individual. As Mackenzie observes, there is a tendency to assume that members of technologically less complex cultures are undifferentiated and subscribe to the communal ideal (1992).

This need to establish generalizations persists, although many researchers now often acknowledge significant behavioral variations in the communities they study. Specific individuals are the basis of cultural variation within a community. They construct alternatives which others may follow. Johnson notes that such cultural risk takers are too often written off as "deviants" or products of the acculturative effects of Western influences on the basis of the presumption that cultural rules apply unambiguously (Johnson 1972). Johnson explains—and we have seen this in the case of the Biase—that traditional farmers do indeed carry standard formulas for dealing with environmental stress. But the rules that define those formulas are not absolute. Adherence to standard cultural survival strategies does not preclude individual inclination to excel or acquire social prestige through apparently aberrant actions that violate standard practices.

Directed Social Change

Ten years ago, Hoben noted that present-day anthropology has contributed to development by undermining these deep-seated assumptions about non-Western people in showing that our behaviors are no less well-thought-out and guided by complex social and cultural factors than the behaviors of Western people are (1982). Later Wright noted that critiques of "directed culture change" studies have enhanced this new focus by stressing how the understanding of existing indigenous institutions can provide the basis for development models, based on a consideration of the environment and the use of local human and natural resources (1988). This new focus has gained greater emphasis, as Escobar observes (1991), with the current requirement by agencies such as USAID that project implementation be preceded by a "social soundness" analysis which assesses the feasibility of development projects and their relevance to the sociocultural environment. Yet economic and ethical problems persist.

Despite the "liberal" stance of some funding agencies, target communities as well as anthropologists must of necessity conform with donor agencies' criteria for implementing specific programs. Anthropologists working in development inevitably have to walk within the familiar but narrow path of satisfying the stipulations of funding agencies and also to make their recommendations locally relevant and culturally acceptable. The efforts to keep to this path continue to draw criticism. Perhaps one of the strongest recently came from Arturo Escobar: "Development anthropology, for all its claim to relevance to local problems, to cultural sensitivity, and to access to interpretive holistic methods, has done no more than recycle, and dress in more localized fabrics, the discourses of modernization and development" (Escobar 1991: 677). This is no doubt true. But despite the garb in which the language for change is adorned, it is of small relevance to the deprived villager if it takes little account of her or his concerns. When near-destitute villagers define their problems and express their need for change, their expression is based on their awareness that neighboring communities are living under less-stressful conditions.

No doubt there are communities in developing countries which seem to strongly resist development projects. But what looks like rejection may be disappointment with development failures that result from faulty assumptions which discount the human personality in its local manifestation. An imperfect fit between projected designs and local reality is inevitable if, as is often the case, research on local needs fails to include the study of what people feel and desire socially and "spiritually," as well as economically (Grayzel 1986). A society that meets its needs without external support or intervention may be less accepting of novel projects. Indeed, this was the case in the early years of British colonial intervention when the self-sufficient Agwagune attacked a British expeditionary force sent to subdue and bring them under colonial rule (Nair 1972). Locals are more aware of the consequences of intervention and skeptical about projects for change, especially if a previous experience was negative. Many projects that target rural communities are beyond the capability of local managerial skills and lack the infrastructural support and expertise needed to sustain them. In such cases, local resistance may result from a previous negative experience. Rural people do not resist change but actively seek to improve their circumstances. Ironically, city elites who are far removed from the rural experience are often more vocal about "protecting" us and "preserving" our cultures than our villagers are. To put it strongly, when rural Africans needed protection from external influences, rationality was submerged in the zeal of "discovery" by missionary soldiers and the need to feed escalating urban populations. It is unreasonable now to have such rural people shoulder the burden of their underdevelopment in the guise of protecting the sanctity of culture.

Poor Biase farmers do not seem to seek the protection that governments and local elites see as a major concern. They take a greater interest in change than their less-poor neighbors who may be concerned with protecting their status (Cancian 1984). While a community's elite may serve as advocates for change to address perceived needs and may influence the direction of change, their representation may not always coincide with the desire of others in the community. Indeed, although African elites have contributed significantly to the discourse on development, it is likely that many rural Africans wish they would keep out of it. For example, G. O. Nwankwo, a Nigerian academic, blames the farmer for Africa's food crisis because he "is a totally frustrated and disillusioned man" who cannot be relied on to feed the nation. "He farms because he has no alternative. He is a tattered, hungry, aging, and dying man with a negative psychology against farming. . . . His overriding ambition is to find all possible ways of getting out of the land and at any rate to ensure that his children and future generations are not subjected to the same fate" (cited in Kiwanuka 1986: 14). Such assertions point to the differential perception that exists between urban elites and their home constituents about reality back in the village. The farmer is indeed often a woman responsible for about 80 percent of domestic food production and filling in two-thirds of agricultural labor time (Longhurst 1988; Green 1989). She is not tattered, aging, and hungry, and her ambition can be measured more closely by her persistence in working the land. Nwankwo's view discounts the dominance of women in farming and finds company in the obsolete 1950s perspective that blamed the poor for complex social problems. This opinion justifies Chambers's assertion that "professional training inculcates an arrogance in which superior knowledge and superior status are assumed. Professionals then see the rural poor as ignorant, backward and primitive, and as people who have only themselves to blame for their poverty" (1986: 6). Coming from an African scholar, Nwankwo's assertion does little to comfort Escobar in his statement, presumably directed at Western scholars, that "development anthropologists, for all their self-proclaimed sensitivity to local conditions, have not escaped the ethnocentricity of the whole development paradigm" (1991: 671).

Often communities desiring socioeconomic change make specific demands to the state, as in the case reported above. These requests are easily ignored because they are often perceived as reflecting no more than an unrealistic desire to replicate the ways of life of the city. In a request the Biase made to their state governor after a flooding disaster in 1970, the governor's response was that their priority should be to rebuild their homes and to get on with their farmwork, although the flood victims had little means to do so.

Nigeria's "Better Life for Rural Women" Program

The governments of some developing countries attempt to respond to rural needs by setting up semiautonomous agencies outside the control of the government. Although such departments are assigned the responsibility for rural development, the government rarely relinquishes enough control to let the agencies achieve their goals. For example, Nigeria's Directorate of Foods, Roads and Rural Infrastructure (DFRRI), set up by the government to improve rural conditions, was discredited by the wife of the then president of the country, who founded an alternative program, the Better Life for Rural Women (BLP). The elitist stance of BLP endeared it very little to some Nigerian women, who said it was wasteful of public funds and only intended to shore up the image of the wives of government officials. Biase women admitted no knowledge of this program, nor any experience whatever of benefits projected to accrue from it. As Aryeetey (1990) observes, the rural people these programs are supposed to help are often illiterate and are not encouraged to participate, much less control these programs. Using examples from Ghana, he attributes the secondary position of rural people in such decision processes both to their ethnic minority status and to their weak local polity. These factors tend to lower their ability to negotiate with governments. Ghana's Upper Regional Agricultural Development Program (URADEP), set up to oversee agricultural development and increase farm incomes, has been hampered by continuous interference by the government, bureaucratic inflexibility, lack of clarity in its objectives, and poor management (Aryeetey 1990: 208).

The implication of programs such as URADEP and BLP for targeting women for development is both sociological and economic. It seems that the public intention of elite city women to alleviate the poor condition of their less-advantaged sisters conflicts profoundly with local realities. The characterization of BLP by other women of high social status "as merely a show of eminent urban women mimicking the rural poor" affirms the position that those directly experiencing disadvantages should be given control of programs aimed at changing their conditions. The lack of respect for women by powerful and powerless men alike is well known. But we also need to clearly articulate how the economic and political asymmetry between urban and rural women can be an obstacle to effectively reaching the "beneficiary" communities.

Obstacles to Development in Biase

A more fruitful approach to bringing about rural economic growth would be to first understand obstacles to change intrinsic to the rural sociocultural milieu and then to assess what change factors could operate within those constraints.

Efforts at fostering economic growth in Biase could start with identifying factors intrinsic to the Biase sociocultural milieu that pose profound obstacles to economic growth. I have suggested the following in the preceding chapters:

1. The Biase physical environment is undergoing problems such as flooding, which affect the equitable distribution of arable land among villagers. These environmental problems intensify the sense of territoriality of some of the historically more powerful communities, such as the Agwagune, and the demand for land rights by communities such as the Adim, which have historically been excluded from land use. The re-

sulting endemic warfare between Biase villages damages the rural economy by curtailing access to resources.

2. The intensification of various environmental troubles and armed aggression has outstripped the effectiveness of indigenous coping methods and encouraged the use of alternative and less functional methods. Surplus production from adoption of high-yielding crops does little to improve the economic circumstances of the Biase because of the lack of adequate storage facilities and the difficulty of getting food quickly to market centers. This problem is made more severe by the periodic marketing system, which constrains distribution of produce, men's lack of participation in the Biase market economy, and the burden placed on women to sustain both the domestic and the communal economies. Men's lack of economic power significantly lowers their authority to enforce social rules or to initiate projects.

3. Although women are the primary entrepreneurs today, they are constrained by cultural factors from maximizing market opportunities.

4. Changes in Biase internal political organization have weakened traditional authority structures, making them less able to respond to pressures exerted by new patterns of social interaction. But the men's tenacious yearning for a once-glorious past remains inflexible to new demands of gender, age, and economic relations and has failed to come to terms with a new configuration in gender and age relations.

5. Since the mystical associations were pivotal for social cohesion, particularly kinship affiliations, their loss of authority left little structure for consolidating kinship obligations. This has led to the diminishing of male authority and the emergence of economic dominance by women.

6. The growing challenge to traditional beliefs and practices has resulted in a loss of awe of and reverence for traditional belief systems. The absence of any visible supernatural retribution as punishment for violating proscriptive rules has demystified the deities and elders and made many community members unwilling to cooperate on locally initiated projects.

But these obstacles to economic growth are not absolute barriers and indeed they exist alongside factors that can be mobilized to foster rural economic growth.

Fostering Change through Cultural Reconfiguration

I have argued so far that utilizing local resources effectively depends largely on understanding culture change clearly and on determining which cultural practices and behaviors are still viable and which are merely vestigial. This argument is not new. Indeed almost thirty years ago, Oliver noted that despite our awareness that all cultures are constantly changing, "our interpretations of cultural transmission are sometimes phrased as though this were a simple matter of passing on an intact model from generation to generation" (1965: 421). And Bromley and Cernea (1989) have been more definite in noting that once-effective indigenous institutions have virtually ceased to exist. As an alternative, they suggest that building rural managerial capacity is central to revitalizing institutional arrangements. "In rural areas, meaningful organizations will need to be legally empowered to take certain actions, and they will then need to formulate the working rules that will define how they propose to function" (1989: 55).

Consequently, we must first determine the extent to which the socioeconomic structure we perceive in the course of doing fieldwork is different from what sustained the community before change became pervasive. This calls not only for familiarity with the ethnographic and theoretical literature but also for formulating questions that elicit the historical patterning of the specific variables under analysis. Such retrospective explanations may provide a historical perspective that will enable researchers and development agencies to better understand the direction and processes of change in the field.

I have already noted how these processes affect Biase sociocultural arrangements. The fact that governments and development agencies sometimes fail to correctly perceive the dynamics of rural change or to assume an inflexibility in socioeconomic conditions (Bromley and Cernea 1989) constrains accurate assessment of locale-specific conditions. The aspects of traditional culture that are most persistent and most strongly resistant to change are the relative powerlessness of women, kinship linkages (as opposed to obligations), and faithfulness to traditional ideological beliefs and practices. Yet even some of these elements have been significantly restructured in many traditional societies, suggesting the need to rethink hypotheses constructed for conditions that no longer exist. I will note these changes again, as they affect the Biase.

Reconfiguring Social Roles

The exclusion of Biase women from political authority as a means of protecting a male-defined status hierarchy does not diminish women's enthusiasm for excelling in the economic sphere or for assuming traditional male responsibilities, even in the face of customary constraints.

The changes reported here about the reconstruction of Biase women's status in the light of diminished male authority and the decrease in men's contribution to household sustenance have increased the incidence of female-headed households, not only among the Biase in southeastern Nigeria but all over Africa. In some areas only one-third of the farm households have a resident male head of household (Green 1989; Alamgir and Arora 1991). Literature on the pivotal role of women in the rural economy is so familiar and commonly accepted that it seems pointless to reiterate it here. But the point keeps recurring because of the consistency with which development programs continue not to let women participate in new projects, and local governments persist in trivializing women's role in food production.

In many rural societies traditional ideologies no longer command fear. It is reported of the Kamba of Kenya that "no matter how sacred an act may be to them, despite all supernatural terrors which one would suppose sufficient to bind them to a common interest, the discordant spirit is yet stronger, and nothing lacks more in their composition than a unanimous feeling" (Dundas, reported in Oliver 1965: 426). The Agwagune have a proscription against engaging in sexual intercourse with the Ugep of Yako because they trace descent to the same apical ancestor as the Agwagune. Over the last twenty years, this proscription has been frequently violated without sanctions. In earlier years violators stayed away from the community in the hope of avoiding punishment. Today, marriage between Ugep and Agwagune is occasionally practiced, although some individuals still oppose it. This suggests that indigenous beliefs and practices among the Biase are neither immutable nor inflexible, and that rural people are neither dogmatic in clinging to traditional beliefs nor unwilling to trade them for more adaptive practices. The characterization of indigenous lifeways as inflexible is likely to hinder emphasis on local participation and may even prevent some rural communities from receiving development assistance. For example, excluding the farmer from the initial phase of project design on the grounds that his traditional beliefs and technology are constraining and "unscientific" (as in the case of the Bakolori Project; chap. 1) also excludes the possibility of mobilizing those psychological factors that might spur the farmer toward greater production.

Kinship linkages are still conceptually important for organizing certain forms of social interaction, such as access and use rights to farmland. But they are not the only means of obtaining such rights, nor does relatedness serve as the primary conduit for the endless interchange of moral and material obligations which supports subsistence activities. Among the Biase, kinship linkages offer little basis for understanding patterns of social interaction, because the role of alternative social groups, such as friendship clubs, age sets, and secular and mystical associations, as critical important alternative cohesive elements has intensified.

This point holds special relevance for farming systems researchers who focus on smallholder farming households in order to better develop technologies on the basis of an intimate understanding of family resources, needs, and the constraints of the current production system (Moock 1986). Variability among households justifies their analytic importance because, as Orlove says, such a perspective more readily admits an examination of the roles of conflict and competition (1980). By observing the significant individual differences in community households, it is possible for different units to be isolated and examined using specific analytic paradigms for understanding the total cultural dynamics.

The analysis of rural decision-making processes continues to present problems for researchers because of the challenges of identifying those individuals responsible for various farming decisions (McKee 1986), and because of the perception of the rural African household as a pooling mechanism that benevolently distributes income to its members (Wong 1984). For example, during my research in Biase, I noticed that variability within households was marked by some individuals maintaining separate cooking pots. People often pointed out that they slept in their family homes but owned separate food pots. This did not preclude their meeting specific obligations to the lineage, such as ritual feasting with ancestral spirits, but it does indicate a significant lack of homogeneity in household organization. Wong suggests that in rural Africa, where the man is often simplistically perceived as being responsible for all decisions, paternal authority should not be uncritically accepted. Decisions in the Biase household may come from any member despite gender, age, or position, depending on the specific requirements of the task.

Polly Hill (1972) stated long ago that agricultural wage labor is absent in the rural areas of West Africa and that whatever wage labor exists is undertaken by subordinate members of the household production unit only after they have met their responsibilities to that unit. The Biase experience does not coincide with Hill's observation. In Biase wage labor is now the primary means of meeting farmwork requirements, with extra labor coming from outside kin (table 9). This is particularly so in cases where the child's formal education in the city has diminished the supply of family labor (see Haswell and Hunt 1991).

An ancillary consequence of education deserves a brief mention here because of its effect on the kinship hierarchy. Younger, more literate kinsmen better able to understand government protocol have emerged as patrilineage heads and village chiefs, shifting authority from the old to the young. This shift of authority from elder men to younger, more literate, lineage members must be recognized to effectively uncover the new subsets of the authority system in African communities.

The patterns of social interaction we now witness in the course of fieldwork differ significantly from the patterns that sustained those societies in the decades before culture change precipitated by intercommunal trade, colonialism, and warfare pervaded those systems. There is structural dysfunction in the coping strategies currently employed by traditional societies, because those strategies are not indigenous; we lack adequate and familiar institutional mechanisms for dealing with alien practices and behaviors we are desperately struggling to make our own. Since such strategies are adjusted only minimally to present realities, rural communities are burdened with the task of sustaining livelihood by means of patterns of social interaction alien to their current environment.

While the changes are irreversible, the mechanisms for coping with them are not. Cernea and Bromley's recognition of this prompts their suggestion that building rural managerial capacity is central to revitalizing institutional arrangements. This goes beyond advocating the participation of community members in project decisions affecting them, by seeking to revitalize and empower existing local institutions to implement and supervise desired social changes. In this process experiences and predispositions should be traded between villagers and project officials, resulting not so much in *implementing* as in *constructing* the final project so that each party significantly affects the outcome (de Alcantara 1991: 610). The neglect of this approach has contributed to the failure of projects in Africa as much as compliance with it has resulted in some of the few success stories.[1]

The most enduring social groups in Biase are the age sets and the

1. By using oral histories and direct observations of events such as the increasing distance women traveled to obtain firewood, villagers in the Koumpentoum Entente project in eastern Senegal assessed the environmental problems they faced from drought and founded and successfully managed a village-based development effort to deal with them (Arnould 1990).

women's associations. These groups strictly maintain order and respect within their ranks, stress accountability by members and leaders, and are unforgiving of infractions. Punishment includes barring members from associating with peers they have grown up with all their lives. Since members are eternally locked into their age sets, everyone strives to maintain good relations with other members in the set as a way of furthering his or her own securities and successes. These circumstances give the leaders of the groups a degree of authority over their members that is respected by all. When problems arise they are more often between age sets than within them, and they are often the result of one age set contesting the authority of another.

Every age set in Agwagune has a male and a female leader who are highly respected by their peers. Along with this esteem, the community vests authority in these leaders and sanctions the exercise of that authority in the achievement of group or community goals. In building local managerial capacity, development agencies should first identify this authority base by talking to community leaders as a recognition of communal civility, and then talking with ordinary community members, who sometimes have a better idea of the political landscape than their leaders. This strategy is often neglected in the familiar temptation to identify the eldest man of the patrilineage as the repository of group allegiance, although current social and political circumstances have resulted in younger men and resourceful women attaining authority positions in many households. Development agencies should work with community persons to identify the authority base, define the task to be accomplished, discover exactly what needs to be done and how it is to be accomplished, set a deadline for completion, and recognize the authority of the association's leader as incontestable by another. The recognition may be in the form of recompense to workers that can only be dispensed by the association's leaders and monitored by the agent.

The leaders of the community and the associations should be responsible for scheduling work, since they know more about local conditions that may impede or foster progress. Locals are also best able to structure a schedule that allows everyone to attend to his personal chores without compromising the task at hand.

The women's associations also operate on the principle of member discipline and respect for the groups' authority. As noted earlier, women frequently announce work and members respond with little hesitation or argument. This disposition has fostered the cooperative ethic among their members and placed their groups in the preeminent position they occupy in Biase. It is important to target them as a durable force for change in Biase by including them at all stages of development planning.

When specific local groups are targeted and given the responsibility to fulfill a particular task, results are often more certain than when accountability is less well defined. Groups so empowered strive to realize the expected goal, partly to give evidence of their resourcefulness to the community and partly to signal their ability in the competition to excel in a communal task. Completed tasks are often symbols of pride for the associations, and members keep constant watch to make sure their projects remain in good condition. Such a project design would strengthen local managerial capacity rather than temporarily improve rural conditions. The local people who organize the project in the first place would also provide the expertise to keep it functioning and be more capable of operating it long after the planners have left.

The major problems the Biase face are flood control, improving transportation, and the development of a monetary exchange economy. The environmental problems may not be easily solved using Biase technology, but the problem-solving mechanisms of Biase social groups encourage a focus on local expertise as a template for solving the problems in the region. Development in Biase does not have to begin with the emergence of a gigantic technological miracle. Rural peoples such as the Biase are content to experience modernity in meaningful but sustainable bits. Such steps could involve only advisory technical experts to direct the locals on where to lay bricks, what strength of rods are required at one location, and how much concrete to pour at another.

The institutions for coordinating this effort are already in place: the age sets, the women's associations, and the various lineages. In spite of the appearance of malaise at the communal level, these local institutions are strong and resolute in enforcing the collective will of their specific groups. As I explained in the cases of the Abu, the Aka-e-Mitin Aneba, and the age sets, the allegiance members pay to their groups sometimes exceeds their allegiance to the community. While this microlevel allegiance may seem to conflict with the general communal interest, it actually binds members in the mutual need to preserve those operational boundaries that define the autonomy of each association. A conflict between two age sets, for instance, arises when one contests the authority of another set it perceives to lack the cohesion and leadership within its group that foster the maintenance of communal boundaries.

Some communities have taken decisive steps to address their flood problems. Odumugom is currently relocating its community some distance away from the flood zone. Relocation will minimize the problem people experience as they move around their communities during the rains, but moving further inland complicates river transportation for coastal communities unless an alternate access to the river is built into the plan. Some communities have reinforced their riverbanks with embankments meant to stem the high tide of the Cross River. This has often not been successful because the materials they use for the projects are easily overwhelmed by the force of the river. The attempts, however, indicate that the Biase are willing to expend time and labor to correct the problems they face; providing them with work materials more durable than sticks, mud, and sand would enable them to apply their labor and their will more profitably.

Biase men say they are not involved in the economy because they lack an all-season road and a good market. Although I have suggested that roads and markets will not necessarily solve the problem, roads are needed to stimulate the economy. The government would benefit from using the premier resource of the Biase, the sand that covers most of its banks in the dry season. The local government can save money in the long term if in the short term it invests money to construct minimally usable roads to some Biase coastal villages to enable this sand to be moved to government construction sites. The cost of such road construction would be quickly absorbed, because money would be saved by not having to buy sand for government construction. To guarantee a return from its investment, the government could have an understanding with the respective villages for control of sand rights over a specified number of years. Villagers would make good income from working on projects involving the transportation of sand to the cities for sale or for construction of business and residential buildings. Men's leisure time may not be productive, but it is a valued component of communal life that could be redirected into such projects. A large-scale project of moving sand might also slow down the rate of accumulation of silt in the riverbed and make surrounding villages less susceptible to floods.

When local structures are set up and activated to involve the locals in the design, implementation, and monitoring of projects, target communities will more easily understand how to harness local skills and resources to meet the demands of development agencies and the expectations they have of each other. Involving the locals in specific projects will create powerful channels for setting and realizing future goals. The enthusiasm will spread to other aspects of the economy and engulf the community in a process of change in which they are essentially the experts.

Conclusion

Reliance on local natural resources is essential to the survival of rural communities that depend fundamentally on farming as a means of livelihood. The need to make such natural resources available at all times or to reduce the risk of their scarcity is at the core of the concern expressed by the United Nation's World Commission on Environment and Development in their 1987 Brundtland Report, "Our Common Future." Sustainable development is the most popular concept to have come out of that report and is defined there as "paths of human progress which meet the needs and aspirations of the present generation without compromising the ability of future generations to meet their needs" (Archibugi and Nijkamp 1990). This global concern with making development sustainable has important implications for discovering ways to better incorporate development activities into rural communities.

First, involving rural communities from the start is not likely to be successful without a clear understanding and incorporation of the community's social organization. Second, target communities are often faulted for the failure of development projects designed around communal organizational principles that have ceased to exist. If projects were designed to strengthen local managerial capacity rather than to temporarily improve rural conditions, rural peoples would be more capable of running operations long after the planners have left. Strengthening rural capacity involves acknowledging the importance of the attitudes, beliefs, and worldview of rural peoples as important instruments that guide them toward supporting and sustaining development, thereby reducing the chance of project failure. Third, the familiar perspectives that have informed implementation of development objectives have produced few results partly because of the emphasis on generalizing standards for development across different ecological zones. If the concept of quality of life, for instance, is tailored to reflect what locals perceive as important, problems of rural destitution will be reduced. Quality of life often may be improved by simple, inexpensive projects that are responsive to rural needs. For example, Nigeria's steel industry does not have to begin functioning at full capacity for rural people to have good drinking water. Participation of community members in implementing the changes affecting them requires articulating their perception of what the problems are; their assessment of how best to address those problems; what local resources they believe could most efficiently be applied to possible solutions; assessing future availability of those resources for continuous support of elements of change; identifying which community members or groups are responsible for operating and maintaining the mechanisms of change; and what powers the community is willing to assign them as they perform those functions.

The local perspective is not necessarily the correct one, but it is a significant source of potential conflict with outside agencies when ignored. The basic strategies for local-level perception and analysis should be to fully discuss the views of local people with them, pointing out problems they have not anticipated, and suggesting alternatives. The Biase, for example, see their problem as a stagnant economy that could be readily corrected if they had a good transportation system and an active marketing network. Our men would be more involved in the local economy if they could take their produce to outside communities rather than limiting their production to meet the needs of local consumption. But they must be made aware of how the changes they desire will affect other aspects of their society, not as a way of discouraging change but as a way of making them more capable of managing the processes of change. Easier access to their villages will improve production, which may in turn energize their economy. But it also will increase the need for land and the potential for conflict between villages. Good roads also provide easier access for enemy villagers to strike quickly and get away before local people are mobilized to counter their assaults. Since many compounds are virtually deserted during peak farming periods, Biase villages will face the possibility of increased urban-style crimes, such as the kidnapping of unsupervised children. Repairing a modern paved roadway will challenge a local technology accustomed to constructing and maintaining trails. Making communities such as the Biase more fully aware of such challenges would enable them to more effectively consider possible alternative development options or to better organize their resources to counter the possible bad consequences of the desired change.

Culture serves as a useful cognitive map for human behavior only because we constantly adjust its components to changing circumstances; culture does not perform that function when it is inflexible to change.[2] Failure to recognize the reconfiguration of indigenous institutions in response to colonialism, to the state, and even to intercommunal relationships is likely to continue to generate inappropriate development goals and less-than-optimal responses by target communities.

2. Hastrup and Elsass observe that culture does not survive through "the conservation of a preconceived identity anchored once and for all in an objectively existing (reified) culture but continuing control by the agents of a particular culture of the shaping of local history" (1990: 306).

Glossary of Agwagune Words

Adimi Mo Aba	"Those pressed to the ground" (former name of the Adim).
Afia	Market.
Aka edong	The house designated for assembly of lineage members.
Aka-e-Mitin Aneba	The women's great meeting.
Aneba (sing. Onegwa)	Women.
Aneba-ogwu	Many wives.
Anerom (sing. Onurom)	Men.
Asanga	A stringed hook suspended from a piece of bamboo, baited, and dropped in fishponds.
Bob	Fishpond.
Ebiabu!	The exclamation that identifies a member of the Abu Association.
Ediba Bob	Fish tax booth.
Egimi	Man-made fishpond.
Egop	Age set.
Ekat Ugom Adim	The entrance into Emomoro used by the Adim when they paid the Agwagune tribute.
Eroi	Bush flooded by the rainy season water.
Eromo	Friends.
Erot Edok	The yam fertility deity.
Etoima	The head of a patrilineage.
Eziba	The earth.
Foofoo	Food processed from cassava or yams.
Gari	Food processed from cassava.
Ibe-e-man	A sacred pond which serves as the home of a fertility deity.
Idom	Rules that prescribe appropriate behavior.

Igot	Village watch.
Igwu	A protective magic that kills anyone who has the intention of committing evil against another.
Ikpafini	Thrift associations.
Ikpo	The designated Agwagune market day.
Ima	Lineage members.
Kobo	One-hundredth of the Nigerian Naira.
Meri-meri	A fishing technique that uses no bait but many hooks strung on a piece of string.
Naira	Monetary unit of Nigeria equal to 100 kobo.
Ofofop	A locally brewed gin.
Okwu Onne	An all-male association of the Emomoro.
Onegwa	Wife.
Onun (pl. Inun)	The head of a political, social, or ritual group.
Onun Erot	The priest of a shrine.
Ovuk	Concubine.
Ugbugba ugim	The new yam festival maiden; a maiden in a fattening room.
Ugim	Fattening the bride.
Ugom Inun	Kingship lineage.
Uso Ekpa Eziba	The place in Emomoro where the founder of Agwagune first built his house.
Uso Onne	Good luck (face).

References

Adams, W. M. 1988. "Rural Protest, Land Policy, and the Planning Process on the Bakolori Project, Nigeria." *Africa* 58 (3): 315–36.

Afigbo, A. E. 1987. *The Igbo and Their Neighbors.* Ibadan: University Press.

Ajayi, Femi, Souleymane Anza, Ben Ephson, Colleen Lowe Morna, and Khelil M. Seck. 1990. "Tools of the Trade: Do Farmers Have the Right Ones?" *African Farmer* 5 (November 1990): 5–13.

Alagoa, E. J. 1971. "The Niger Delta States and Their Neighbors to 1800." In M. Crowder and J. F. A. Ajayi, eds., *History of West Africa,* vol. 1. London: Longman.

Alamgir, Mohiuddin, and Poonam Arora. 1991. *Providing Food Security for All.* International Fund for Agricultural Development. New York: New York University Press.

Andretta, Elizabeth H. 1989. "Symbolic Continuity, Material Discontinuity, and Ethnic Identity among Murle Communities in the Southern Sudan." *Ethnology* 28 (1): 17–31.

Anikpo, Mark. 1986. "The Place of Traditional Rituals and Symbolic Expressions in Nigerian Agriculture: The Case of Igboland. Pp. 95–111 in Adefolu Akinbode, Bryan Stoten, and Rex Ugorji, eds., *The Role of Traditional Rulers and Local Governments in Nigerian Agriculture.* Ilorin, Nigeria: Agricultural and Rural Management Training Institute (ARMTI).

Anyatonwu, G. N. 1986. "An Appraisal of the Federal Government Policy and Contributions to Agricultural Development in Nigeria". Pp. 55–73 in Adeniyi Osuntogun and Rex Ugorji, eds., *Financing Agricultural Development in Nigeria.* Ilorin, Nigeria: Atoto Press.

Archibugi, F., and P. Nijkamp, eds. 1990. *Economy and Ecology: Towards Sustainable Development.* Dordrecht, Netherlands: Kluwer Academic Publishers.

Arnould, Eric J. 1990. "Changing the Terms of Rural Development: Collaborative Research in Cultural Ecology in the Sahel." *Human Organization* 49 (4): 339–54.

Aryeetey, Ernest. 1990. "Decentralization for Rural Development: Exogenous Factors and Semi-Autonomous Program Units in Ghana" *Community Development Journal* 25 (3): 206–14.

Attoe, Stella. 1990. *A Federation of the Biase People.* Enugu, Nigeria: Harris Publishers.

Barber, W. J. 1960. "Economic Rationality and Behavior Patterns in an Underdeveloped Area: A Case Study of African Economic Behavior in the Rhodesias." *Economic Development and Cultural Change* 8 (3): 237–51.

Barker, Jonathan. 1982. "Thinking about the Peasantry." *Canadian Journal of African Studies* 16 (3): 601–5.

Barlett, Peggy F. 1982. *Agricultural Choice and Change: Decision-Making in a Costa Rican Community.* New Brunswick: Rutgers University Press.

Barrows, Richard, and Michael Roth. 1990. "Land Tenure and Investment in African Agriculture: Theory and Evidence." *Journal of Modern African Studies* 28 (2): 265–97.

Barry, H., III, I. L. Child, and M. K. Bacon. 1959. "Relation of Child Training to Subsistence Economy." *American Anthropologist* 61: 51–63.

Barton, R. F. 1922. *Ifugao Economics.* University of California Publications in American Archaeology and Ethnology 15: 385–446.

Basden, G. T. 1966. *Among the Ibos of Nigeria.* London: Frank Cass.

Batalla, Guillermo B. 1966. "Conservative Thought in Applied Anthropology: A Critique." *Human Organization* 25: 89–92.

Bates, Robert H. 1990. "Capital, Kinship, and Conflict: The Structuring Influence of Capital in Kinship Societies." *Canadian Journal of African Studies* 24 (2): 151–64.

Belloncle, Guy. 1973. "Listening to the Peasant." *Ceres,* May–June, 1973.

Bendix, R. 1967. "Tradition and Modernity Reconsidered." *Comparative Studies of Society and History* 9: 292–346.

Bentley, Jeffery W. 1987. "Economic and Ecological Approaches to Land Fragmentation: In Defense of a Much Maligned Phenomenon." *Annual Review of Anthropology* 16: 31–67.

Bernard, H. Russell. 1988. *Research Methods in Cultural Anthropology.* Newbury Park, California: Sage Publications.

Berry, Sara. 1989. "Social Institutions and Access to Resources." *Africa* 59:41–55.

Blaikie, Piers. 1989. "Environment and Access to Resources in Africa." *Africa* 59 (1): 18–40.

Bodley, John H. 1985. *Anthropology and Contemporary Human Problems.* 2d ed. Palo Alto, California: Mayfield.

———. 1990. *Victims of Progress.* 3d ed. Mountain View, California: Mayfield.

Boyle, Walden Philip. 1977. "Contract and Kinship: The Economic Organization of the Beni Mguild Berbers of Morocco." Ph.D diss., UCLA.

Brokensha, David, and Jack Glazier. 1973. "Land Reform among the Mbeere of Central Kenya." Africa 43 (3): 182–206.

Bromley, Daniel W., and Michael M. Cernea. 1989. *The Management of Common Property Natural Resources: Some Conceptual and Operational Fallacies.* Washington, D.C.: World Bank Discussion Papers 57.

Brosius, J. Peter. 1988. "Significance and Social Being in Ifugao Agricultural Production." *Ethnology* 27 (1): 97–110.

Buller, Henry, and Susan Wright. 1990. *Rural Development: Problems and Practices.* Brookfield, Vermont: Gower.

Byres, Terry, and Ben Crow. 1988. "New Technology and New Masters for the Indian Countryside." Pp. 163–81 in Ben Crow et al., eds., *Survival and Change in the Third World.* New York: Oxford University Press.

Cancian, Frank, 1984. "Risk and Uncertainty in Agricultural Decision Making." Pp. 161–76 in Peggy Barlett, ed., *Agricultural Decision Making: Anthropological Contributions to Rural Development.* London: Academic Press.
Cannon, Terry. 1990. *Rural People, Vulnerability, and Flood Disasters in the Third World.* The Hague: Institute of Social Studies Working Papers.
Caplan, Pat. 1989. "Perceptions of Gender Stratification." *Africa* 59 (2): 196–208.
Carney, Judith, and Michael Watts. 1990. "Manufacturing Dissent: Work, Gender, and the Politics of Meaning in a Peasant Society." *Africa* 60 (2):207–41.
Chambers, Robert. 1986. *Rural Development: Putting the Last First.* London: Longman Group.
Chibnik, Michael, and Wil de Jong. 1989. "Agricultural Labor Organization in Ribereno Communities of the Peruvian Amazon." *Ethnology* 28 (1): 75–95.
Chick, Gary E. 1984. "Leisure and the Development of Culture." *Annals of Tourism Research* 11: 623–26.
———. 1986a. "Leisure, Labor, and the Complexity of Culture: An Anthropological Perspective." *Journal of Leisure Research* 18 (3): 154–68.
———. 1986b. Review of *Leisure and Recreation Concepts: A Critical Analysis,* by Jay S. Shivers. *Journal of Leisure Research* 18 (4): 337–39.
Collier, Paul, Samir Radwan, and Samuel Wangwe, with Albert Wagner. 1986. *Labor and Poverty in Rural Tanzania: Ujamaa and Rural Development in the United Republic of Tanzania.* Oxford: Clarendon Press.
Crow, Ben, et al. 1988. *Survival and Change in the Third World.* New York: Oxford University Press.
Dannhaeuser, N. 1987. "Marketing Systems and Rural Development: A Review of Consumer Goods Distribution." *Human Organization* 46 (2): 177–85.
de Alcantara, Cynthia Hewitt. 1991. Review of *Encounters at the Interface: A Perspective on Social Discontinuities in Rural Development,* by Norman Long. *Development and Change* 22 (3): 609–10.
Diamond, Larry. 1988. "Nigeria: Pluralism, Statism, and the Struggle for Democracy." Pp. 33–91 in Larry Diamond, Juan J. Linz, and Seymour Martin Lipset, eds., *Democracy in Developing Countries: Africa,* vol. 2. Boulder, Colorado: Lynne Rienner Publishers.
Donham, Donald L. 1985. *Work and Power in Maale, Ethiopia.* Ann Arbor, Michigan: UMI Research Press.
Donnelly, Jack. 1984. "Human Rights and Development: Complementary or Competing Concerns?" *World Politics* 36 (2): 255–83
Erasmus, C. 1955. "Culture Structure and Process: The Occurrence and Disappearance of Reciprocal Farm Labor." *Southwestern Journal of Anthropology.* 12: 444–69.
Escobar, Arturo. 1991. "Anthropology and the Development Encounter: The Making and Marketing of Development Anthropology." *American Ethnologist* 18 (4): 658–82.
Fagerlund, V. G, and R. H. T. Smith. 1972. "A Preliminary Map of Market Periodicities in Ghana." *Journal of Developing Areas.* 4: 333–74.
Farmer, B. H. 1960. "On Not Controlling Subdivision in Paddy Lands." *Transactions of the Institute of British Geographers* 28, 225–35.

Ferguson, R. Brian. 1989. "Ecological Consequences of Amazonian Warfare." *Ethnology* 28 (3): 249–64.

Finan, Timothy J. 1988. "Market Relations and Market Performance in Northeast Brazil." *American Ethnologist* 15 (4): 694–709.

Finsterbusch, Kurt, and Warren A. Van Wicklin III. 1989. "Beneficiary Participation in Development Projects: Empirical Tests of Popular Theories." *Economic Development and Cultural Change* 37 (3): 573–93.

Food and Agriculture Organization of the United Nations. *African Agriculture: The Next 25 Years.* FAO Report. Cited in *West Africa,* January 30–February 5, 1989, p. 136.

Forde, C. Daryll. 1939. "Government in Umor." *Africa* 12 (2): 129–61.

———. 1956. *Efik Traders of Old Calabar.* London: Oxford University Press.

———. 1964. *Yako Studies.* Ibadan, Nigeria: Oxford University Press.

Foster, B. L. 1978. "Ethnicity and Commerce." *American Ethnologist* 1: 437–48.

Foster, George M. 1965. "Peasant Society and the Image of Limited Good." *American Anthropologist* 67 (2): 293–315.

Frank, Barbara. 1990. "From Village Autonomy to Modern Village Administration under the Kulere of Central Nigeria." *Africa* 60 (2): 270–93.

Fruzzetti, Lina, and Akos Ostor. 1990. *Culture and Change along the Blue Nile: Courts, Markets, and Strategies for Development.* Boulder, Colorado: Westview Press.

Gill, Gerard J. 1991. *Seasonality and Agriculture in the Developing World: A Problem of the Poor and Powerless.* New York: Cambridge University Press.

Gladwin, Christina H., and Della McMillan. 1989. "Is a Turnaround in Africa Possible without Helping Women to Farm?" *Economic Development and Cultural Change* 37 (2): 345–69.

Goldschmidt, Walter. 1971. "Introduction: The Theory of Cultural Adaptation." Pp. 1–22 in Robert B. Edgerton, ed., *The Individual in Cultural Adaptation.* Berkeley: University of California Press.

———. 1990. *The Human Career: The Self in the Symbolic World.* Oxford: Oxford University Press.

Grayzel, John. 1986. "Libido and Development: The Importance of Emotions in Development Work." Pp. 147–65 in Michael M. Horowitz and Thomas Painter, eds., *Anthropology and Rural Development in West Africa.* Boulder, Colorado: Westview Press.

Green, Gary P. 1985. "Large-Scale Farming and the Quality of Life in Rural Communities: Further Specification of the Goldschmidt Hypothesis." *Rural Sociology* 50 (2): 262–74.

Green, Reginald Herbold. 1989. *Degradation of Rural Development: Development of Rural Degradation—Change and Peasants in Sub-Saharan Africa.* Institute of Development Studies Discussion Paper 265.

Greenwood, D. J. 1976. *Unrewarding Wealth: The Commercialization and Collapse of Agriculture in a Spanish Basque Town.* New York: Cambridge University Press.

Hammond, Peter B. 1959. "Economic Change and Mossi Acculturation." Pp. 238–56 in William R. Bascom and Melville J. Herskovits, eds., *Continuity and Change in African Cultures.* Chicago: University of Chicago Press.

Harris, Marvin. 1959. "The Economy Has No Surplus?" *American Anthropologist* 61: 185–99.

———. 1981. *Why Nothing Works: The Anthropology of Daily Life*. New York: Simon and Schuster.

———. 1987. "India's Sacred Cow." Pp. 208–19 in Barbara Spradley and Carolyn McCurdy, eds., *Conformity and Conflict*. Boston: Little, Brown.

Harris, Rosemary. 1965. *The Political Organization of the Mbembe, Nigeria*. London: Her Majesty's Stationery Office.

Hart, Keith. 1979. *The Development of Commercial Agriculture in West Africa*. Discussion Paper for the U.S. Agency for International Development.

Hastrup, Kirsten, and Peter Elsass. 1990. "Anthropological Advocacy: A Contradiction in Terms." *Current Anthropology* 31 (3): 301–11.

Haswell, Margaret, and Diana Hunt, eds. 1991. *Rural Households in Emerging Societies: Technology and Change in Sub-Saharan Africa*. Providence, Rhode Island: Berg.

Hatch, Elvin. 1973. *Theories of Man and Culture*. New York: Columbia University Press.

———. 1987. "The Cultural Evaluation of Wealth: An Agrarian Case Study." *Ethnology* 26 (1): 37–50.

———. 1989. "Theories of Social Honor." *American Anthropologist* 91 (2): 341–53.

Haugerud, Angelique. 1989. "Land Tenure and Agrarian Change in Kenya." *Africa* 59 (1): 61–90.

Hecht, Susanna, and Alexander Cockburn. 1990. *The Fate of the Forest: Developers, Destroyers, and Defenders of the Amazon*. New York: Harper Collins.

Hill, K., H. Kaplan, K. Hawkes, and A. M. Hurtado. 1985. "Men's Time Allocation to Subsistence Work among the Ache of Eastern Paraguay." *Human Ecology* 13: 29–47.

Hill, Polly. 1963. "Markets in Africa: A Review." *Journal of Modern African Studies* 1 (4): 441–53.

———. 1972. *Rural Hausa*. Cambridge: Cambridge University Press.

Hoben, Allan. 1982. "Anthropologists and Development." *Annual Review of Anthropology* 11: 349–75.

———. 1984. "Agricultural Decision Making in Foreign Assistance: An Anthropological Analysis." Pp. 337–69 in Peggy F. Barlett, ed., *Agricultural Decision Making: Anthropological Contributions to Rural Development*. Orlando, Florida: Academic Press.

Hollos, Marida. 1991. "Migration, Education, and the Status of Women in Southern Nigeria." *American Anthropologist* 93 (4): 852–70.

Huizer, Gerrit. 1970. "'Resistance to Change' and Radical Peasant Mobilization: Foster and Erasmus Reconsidered." *Human Organization* 29 (4): 303–12.

Hyden, Goran. 1980. *Beyond Ujamaa in Tanzania: Underdevelopment and an Uncaptured Peasantry*. Berkeley: University of California Press.

Ingold, Tim. 1986. *The Appropriation of Nature: Essays on Human Ecology and Social Relations*. Manchester: Manchester University Press.

Iniama, Ededet Akpan. 1983. "The Impact of the Calabar-Ikom Highway on Agricultural Development in the Cross River State of Nigeria." Ph.D. diss., University of Iowa.

Ives, Jack D., and Bruno Messerli. 1989. *The Himalayan Dilemma: Reconciling Development and Conservation.* New York: Routledge.

Jack, D. R. L. 1988. "Community Development in the Process of Nation Building in Africa." Pp. 1–12 in Amechi Nweze, ed., *Perspectives on Community and Rural Development in Nigeria.* Jos, Nigeria: Center for Development Studies.

Jennings, J. H., and S. O. Oduah. 1966. *A Geography of the Eastern Provinces of Nigeria.* Ibadan, Nigeria: Cambridge University Press.

Johnson, Allen W. 1972. "Individuality and Experimentation in Traditional Agriculture." *Human Ecology* 1 (2): 149–59.

———. 1978. *Quantification in Cultural Anthropology: An Introduction to Research Design.* Stanford: Stanford University Press.

Johnson, Hazel, and Ben Crow. 1988. "Developing Production on the Land." Pp. 127–146 in Ben Crow et al., eds., *Survival and Change in the Third World.* New York: Oxford University Press.

Kanu, Mary. 1990. *The African Guardian.*

Karp, Ivan, 1986, "African Systems of Thought." Pp. 199–211 in Phyllis M. Martin and Patrick O'Meara, eds. *Africa* 2d. ed. Bloomington: Indiana University Press.

Keesing, Roger. 1975. *Kin Groups and Social Structure.* New York: Holt, Rinehart and Winston.

Kim, Choong Soon. 1990. "The Role of the Non-Western Anthropologist Reconsidered: Illusion versus Reality." *Current Anthropology* 31 (2): 196–201.

Kiwanuka, Semakula. 1986. "Feeding the Nation: What Are the Alternative Strategies?" Pp. 9–33 in Adeniyi Osuntogun and Rex Ugorji, eds., *Financing Agricultural Development in Nigeria.* Ilorin, Nigeria: Atoto Press.

Knack, Martha C. 1989. "Contemporary Southern Paiute Women and the Measurement of Women's Economic and Political Status." *Ethnology* 28 (3): 233–48.

Knight, C. Gregory. 1974. Ecology and Change: Rural Modernization in an African Community. New York: Academic Press.

Kottak, Conrad Phillip. 1991. *Cultural Anthropology.* New York: McGraw-Hill.

Lee, Richard. 1984. *The Dobe !Kung.* New York: Holt, Rinehart and Winston.

Le Guennec-Coppens, Françoise. 1989. "Social and Cultural Integration: A Case Study of the East African Hadramis." *Africa* 59 (2): 185–95.

Levine, Nancy E. 1988. "Women's Work and Infant Feeding: A Case from Rural Nepal." *Ethnology* 27 (3): 231–51.

Lindskog, Per, and Jan Lundqvist. 1989. *Why Poor Children Stay Sick: The Human Ecology of Child Health and Welfare in Rural Malawi.* Research Report no. 85. Uppsala: Scandinavian Institute of African Studies.

Lipton, Michael. 1985. *Land Assets and Rural Poverty.* World Bank Staff Working Papers no. 744. Washington, D.C.: World Bank.

Lofchie, M. F. 1978. "African Crisis and Economic Liberalization in Tanzania." *Journal of Modern African Studies* 16 (3): 468–69.

Longhurst, Richard. 1988. "Policy Approaches towards Small Farmers." Pp. 183–96 in Giovanni Andrea Cornia, Richard Jolly, and Frances Stewart, eds., *Adjustment with a Human Face.* New York: Oxford University Press.

Low, Allan. 1986. "On-Farm Research and Household Economics." Pp. 71–91 in Joyce Lewinger Moock, eds., *Understanding Africa's Rural Households and Farming Systems.* Boulder, Colorado: Westview Press.

MacDonald, Andrew S. 1989. *Nowhere to Go but Down: Peasant Farming and the International Development Game.* London: Unwin Hyman.

Mackenzie, Fiona. 1992. "Development from Within? The Struggle to Survive." Pp. 1–32 in D. R. Fraser Taylor and Fiona Mackenzie, eds., *Development from Within: Survival in Rural Africa.* New York: Routledge.

Magubane, B., and J. G. Faris. 1985. "On the Political Relevance of Anthropology." *Dialectical Anthropology* 9 (1–4): 91–104.

Martin, Susan. 1988. *Palm Oil and Protest: An Economic History of the Ngwa Region, South-Eastern Nigeria, 1800–1980.* Cambridge: Cambridge University Press.

Mason, John P. 1990. "An Anthropologist's Contribution to Libya's National Human Settlement Plan." Pp. 155–74 in Muneera Salem-Murdock and Michael M. Horowitz, eds., with Monica Sella, *Anthropology and Development in North Africa and the Middle East.* Boulder, Colorado: Westview Press.

McFarlan, Donald M. 1957. *Calabar: The Church of Scotland Mission,* founded 1846. New York: Thomas Nelson and Sons.

McKee, Katharine. 1986. "Household Analysis as an Aid to Farming Systems Research." Pp. 188–98 in Joyce Lewinger Moock, ed., *Understanding Africa's Rural Households and Farming Systems.* Boulder, Colorado: Westview Press.

McKone, C. E. 1985. "Food Production As If People Matter." *International Cooperative Information* 78 (1): 23–31.

Mellor, John W. 1969. "The Subsistence Farmer in Traditional Economies." Pp. 209–28 in Clifton R. Wharton Jr., ed., *Subsistence Agriculture and Economic Development.* Chicago: Aldine.

Mencher, Joan P. 1988. "Women's Work and Poverty: Women's Contribution to Household Maintenance in South India." Pp. 99–119 in Daisy Dwyer and Judith Bruce, eds., *A Home Divided: Women and Income in the Third World.* Stanford: Stanford University Press.

Meneses, Eloise Hiebert. 1987. "Traders and Marginality in a Complex Social System." *Ethnology* 26 (4): 231–44.

Middleton, John. 1992. *The World of the Swahili: An African Mercantile Tradition.* New Haven: Yale University Press.

Mintz, S. 1971. "Men, Women, and Trade." *Comparative Studies in Society and History* 13:247–69.

Montgomery, John D. 1990. "How Facts Replace Fads: Social Science and Social Development." *Comparative Politics* 22 (2): 237–48.

Moock, Joyce Lewinger. 1986. Pp. 1–10 in Joyce Lewinger Moock, ed., *Understanding Africa's Rural Households and Farming Systems.* Boulder, Colorado: Westview Press.

Moore, M. P. 1975. "Co-operative Labor in Peasant Agriculture." *Journal of Peasant Studies* 2: 270–91.

Moran, Mary H. 1988. "Woman and 'Civilization': The Intersection of Gender and Prestige in Southeastern Liberia." *Canadian Journal of African Studies* 22 (3): 491–501.

Murdock, George P. 1959. *Africa: Its Peoples and Their Culture*. London: McGraw Hill.

Myers, Norman. 1984. The Primary Source: Tropical Forests and Our Future. New York: W. W. Norton.

Myrdal, Gunnar. 1968. *Asian Drama: An Inquiry into the Poverty of Nations*, vol. 3: 2121–38. New York: Pantheon Books.

Nair, Kannan K. 1972. *Politics and Society in South Eastern Nigeria 1841–1906*. London: Frank Cass.

Nardi, Bonnie A. 1983. "Goals in Reproductive Decision Making." *American Ethnologist* 10 (4): 697–714.

Nash, Manning. 1971. "Market and Indian Peasant Economies." Pp. 161–77 in Teodor Shanin, ed., *Peasants and Peasant Societies*. Baltimore: Penguin.

Neale, Walter C., and John B. Viar. 1989. Review of *Development Economics on Trial: The Anthropological Case for the Prosecution*, by Polly Hill. In *Economic Development and Cultural Change* 37 (4): 861–64.

Nwankwo, G. O. 1981 "Agricultural Finance Policy and Strategy in the 80s." *Agricultural Credit and Finance in Nigeria: Problems and Prospects*, a seminar organized by the Central Bank of Nigeria, Lagos.

Nyerges, A. Endre. 1987. "The Development Potential of the Guinea Savanna: Social and Ecological Constraints in the West African 'Middle Belt.'" Pp. 316–36 in Peter Little and Michael M. Horowitz, with A. Endre Nyerges, eds., *Lands at Risk in the Third World: Local-Level Perspectives*. Boulder, Colorado: Westview Press.

Okoli, F. C. 1989. "Rural Development in Nigeria: Another Battle Ground for Predators?" Pp. 42–57 in L. A. Ega, T. K. Atala, and J. M. Baba, eds., *Developing Rural Nigeria: Problems and Prospects*. Nigerian Rural Sociological Association.

Oliver, Symmes C. 1965. "Individuality, Freedom of Choice, and Cultural Flexibility of the Kamba." *American Anthropologist* 67 (2): 421–28

Orlove, Benjamin S. 1980. "Ecological Anthropology." *Annual Review of Anthropology* 9: 235–73.

Ottenberg, Simon. 1968. *Double Descent in an African Society: The Afikpo Village Group*. Seattle: University of Washington Press.

Painter, Thomas M. 1987. "Bringing Land Back In: Changing Strategies to Improve Agricultural Production in the West African Sahel." Pp. 144–63 in Peter D. Little and Michael M Horowitz, with A. Endre Nyerges, eds., *Lands at Risk in the Third World: Local-Level Perspectives*. Boulder, Colorado: Westview Press.

Parker, Barbara. 1988. "Moral Economy, Political Economy, and the Culture of Entrepreneurship in Highland Nepal." *Ethnology* 27 (2): 181–94.

Partidge, William L., and Elizabeth M. Eddy. 1978. "The Development of Applied Anthropology in America." Pp. 3–45 in Elizabeth M. Eddy and William L. Partridge, eds., *Applied Anthropology in America*. New York: Columbia University Press.

Pelzer, K. J. 1945. *Pioneer Settlement in the Asiatic Tropics*. New York: American Geographical Society.

Pingali, Prabhu, Yves Bigot, and Hans Binswanger. 1987. *Agricultural Mechaniza-*

tion and the Evolution of Farming in Sub-Saharan Africa. Washington, D.C.: World Bank; Baltimore: Johns Hopkins University Press.

Pitt, David C. 1976. *Development from Below: Anthropologists and Development Situations.* The Hague and Paris: Mouton.

Polanyi, Karl, Conrad M. Arensberg, and Harry W. Pearson. 1957. *Trade and Market in the Early Empires: Economies in History and Theory.* Glencoe, Illinois: Free Press.

Ramirez-Faria, Carlos. 1991. *The Origins of Economic Inequality between Nations.* Cambridge, Massachusetts: Unwin Hyman.

Rogers, Everett M. 1969. *Modernization among Peasants.* New York: Holt, Rinehart and Winston.

Ruttan, Vernon W. 1988. "Cultural Endowments and Economic Development: What Can We Learn from Anthropology?" *Economic Development and Cultural Change* 36 (3) (Supplement): S247–71.

Safilios-Rothschild, Constantina. 1988. The Impact of Agrarian Reform on Women's Incomes in Rural Honduras." Pp. 216–28 in Daisy Dwyer and Judith Bruce, eds., *A Home Divided: Women and Income in the Third World.* Stanford: Stanford University Press.

Sanday, P. 1974. "Female Status in the Public Domain." Pp. 189–206 in M. Rosaldo and L. Lamphere, eds., *Woman, Culture, and Society.* Stanford: Stanford University Press.

Schneider, Harold. 1970. *The Wahi Wanyaturu: Economics in an African Society.* Chicago: Aldine.

Schultz, Theodore. 1964. *Transforming Traditional Agriculture.* New Haven: Yale University Press.

Shiawoya, Emman L. 1986. "Small-scale Farmers, Local Governments, and Traditional Rulers in Agricultural Production." Pp. 45–66 in Adefolu Akinbode, Bryan Stoten, and Rex Ugorji, eds., *The Rolel of Traditional Rulers and Local Governments in Nigerian Agriculture.* Ilorin, Nigeria: The Agricultural and Rural Management Training Institute (ARMTI).

Shipton, Parker. 1990. "African Famines and Food Security: Anthropological Perspectives." *Annual Review of Anthropology* 19:353–94.

Shivers, Jay. 1981. *Leisure and Recreation Concepts: A Critical Analysis.* Boston: Allyn and Bacon.

Silberfein, Marilyn. 1989. *Rural Change in Machakos, Kenya: A Historical Geography Perspective.* New York: University Press of America.

Sklar, Richard L. 1991. "The Contribution of Tribalism to Nationalism in Western Nigeria." Pp. 13–22 in Richard L. Sklar and C. S. Whitaker, eds., *African Politics and Problems in Development.* Boulder, Colorado: Lynne Rienner.

Smith, Robert H. T. 1978. "Periodic Market-Places, Periodic Marketing, and Travelling Traders." Pp. 11–30 in Robert H. T. Smith, ed., *Market-Place Trade—Periodic Markets, Hawkers, and Traders in Africa, Asia, and Latin America.* Vancouver, Canada: Center for Transportation Studies.

Swetnam, John J. 1988. "Women and Markets: A Problem in the Assessment of Sexual Inequality." *Ethnology* 27 (4): 327–38.

Swindell, K., and A. B. Mamman. 1990. "Land Expropriation and Accumulation in the Sokoto Periphery, Northwest Nigeria: 1976–1986." *Africa* 60 (2): 173–86.

Talbot, P. Amaury. 1960. *The Peoples of Southern Nigeria,* vol. 1. London:; Oxford University Press.

Turrittin, Jane, 1988. "Men, Women, and Market Trade in Rural Mali, West Africa." *Canadian Journal of African Studies* 22 (3): 583–604.

Ubi, Otu A. 1986. "Economic and Cultural Relations between Agwagune and Efik in Old Calabar 1800–1900." Paper presented at the International Seminar on Old Calabar History, July 29–31, 1986.

Uyanga, Joseph T. 1980. *A Geography of Rural Development in Nigeria.* Washington, D.C.: University Press of America.

Uzozie, L. C. 1981. "The Changing Context of Land Use Decisions: Three Family Farms in the Yam Cultivation Zone of Eastern Nigeria, 1964–1977." *Africa* 51 (2): 678–93.

Vaughan, James H. 1986. "Population and Social Change." Pp. 160–80 in Phyllis M. Martin and Patrick O'Meara, eds., *Africa.* 2d ed. Bloomington: Indiana University Press.

Vayda, Andrew P. 1969. "Expansion and Warfare among Swidden Agriculturalists." Pp. 202–20 in Andrew P. Vayda, ed., *Environment and Cultural Behavior: Ecological Studies in Cultural Anthropology.* Garden City, New York: Natural History Press.

von Braun, Joachim, and Patrick J. R. Webb. 1989. "The Impact of New Crop Technology on the Agricultural Division of Labor in a West African Setting." *Economic Development and Cultural Change* 37 (3): 513–34.

Wade, Nicholas. 1974. "Green Revolution: Problems of Adopting New Technology." *Science,* December 20 and 27.

Webster, Gary S. 1990. "Labor Control and Emergent Stratification in Prehistoric Europe." *Current Anthropology* 31 (4): 337–66.

Wessing, Robert. 1987. "Electing a Lurah in West Java, Indonesia: Stability and Change." *Ethnology* 26 (3): 165–78.

Wey, S. O. 1988. "Community Development and the Process of Economic Development." Pp. 39–48 in Amechi Nweze, ed., *Perspectives on Community and Rural Development.* Jos, Nigeria: Center for Development Studies.

White, Leslie A. 1959. *The Evolution of Culture: The Development of Civilization to the Fall of Rome.* New York: McGraw-Hill.

Wilson-Moore, M. 1989. "Women's Work in Homestead Gardens: Subsistence, Patriarchy, and Status in Northwest Bangladesh." *Urban Anthropology* 18 (3–4): 281–98.

Wong, Diana. 1984. "The Limits of Using the Household as a Unit of Analysis." Pp. 56–63 in Joan Smith, Immanuel Wallerstein, and Hans-Dieter Evers, eds., *Households and the World Economy.* Beverly Hills: Sage.

Wright, Robin M. 1988. "Anthropological Presuppositions of Indigenous Advocacy." *Annual Review of Anthropology* 17: 365–90.

Index